THE ASTRONOMY
OF THE BIBLE

From the Painting by Sir Edward Burne-Jones in the Birmingham Art Gallery.

THE STAR OF BETHLEHEM.

"We have seen His star in the east, and are come to worship Him."

[Frontispiece.

THE ASTRONOMY OF THE BIBLE

AN ELEMENTARY COMMENTARY ON THE
ASTRONOMICAL REFERENCES
OF HOLY SCRIPTURE

BY

E. WALTER MAUNDER, F.R.A.S.

AUTHOR OF
'THE ROYAL OBSERVATORY, GREENWICH: ITS HISTORY AND WORK,'
AND 'ASTRONOMY WITHOUT A TELESCOPE'

WITH THIRTY-FOUR ILLUSTRATIONS

WILDSIDE PRESS

To
MY WIFE
My helper in this Book

and in all things.

PREFACE

WHY should an astronomer write a commentary on the Bible?

Because commentators as a rule are not astronomers, and therefore either pass over the astronomical allusions of Scripture in silence, or else annotate them in a way which, from a scientific point of view, leaves much to be desired.

Astronomical allusions in the Bible, direct and indirect, are not few in number, and, in order to bring out their full significance, need to be treated astronomically. Astronomy further gives us the power of placing ourselves to some degree in the position of the patriarchs and prophets of old. We know that the same sun and moon, stars and planets, shine upon us as shone upon Abraham and Moses, David and Isaiah. We can, if we will, see the unchanging heavens with their eyes, and understand their attitude towards them.

It is worth while for us so to do. For the immense advances in science, made since the Canon of Holy Scripture was closed, and especially during the last three hundred years, may enable us to realize the significance of a most remarkable fact. Even in those early ages, when to all the nations surrounding Israel the heavenly bodies were objects for divination or idolatry, the attitude of the sacred writers toward them was perfect in its sanity and truth.

Astronomy has a yet further part to play in Biblical study. The dating of the several books of the Bible, and the relation of certain heathen mythologies to the Scripture narratives of the world's earliest ages, have received much attention of late years. Literary analysis has thrown much light on these subjects, but hitherto any evidence that astronomy could give has been almost wholly neglected; although, from the nature of the case, such evidence, so far as it is available, must be most decisive and exact.

I have endeavoured, in the present book, to make an astronomical commentary on the Bible, in a manner that shall be both clear and interesting to the general reader, dispensing as far as possible with astronomical technicalities, since the principles concerned are, for the most part, quite simple. I trust, also, that I have taken the first step in a new inquiry which promises to give results of no small importance.

E. WALTER MAUNDER.

St. John's, London, S.E.
January 1908.

CONTENTS

BOOK I
THE HEAVENLY BODIES

CHAPTER I. THE HEBREW AND ASTRONOMY

Modern Astronomy—Astronomy in the Classical Age—The Canon of Holy Scripture closed before the Classical Age—Character of the Scriptural References to the Heavenly Bodies—Tradition of Solomon's Eminence in Science—Attitude towards Nature of the Sacred Writers —Plan of the Book 3

CHAPTER II. THE CREATION

Indian Eclipse of 1898—Contrast between the Heathen and Scientific Attitudes—The Law of Causality—Inconsistent with Polytheism—Faith in One God the Source to the Hebrews of Intellectual Freedom—The First Words of Genesis the Charter of the Physical Sciences—The Limitations of Science—"Explanations" of the First Chapter of Genesis—Its Real Purposes—The Sabbath . 12

CHAPTER III. THE DEEP

Babylonian Creation Myth—Tiamat, the Dragon of Chaos— Overcome by Merodach—Similarity to the Scandinavian Myth—No Resemblance to the Narrative in Genesis— Meanings of the Hebrew Word *tehom*—Date of the Babylonian Creation Story 25

CHAPTER IV. THE FIRMAMENT

Twofold Application of the Hebrew Word *raqia'*—Its Etymological Meaning—The Idea of Solidity introduced by the "Seventy"—Not the Hebrew Idea—The "Foundations" of Heaven and Earth—The "Canopy" of Heaven—The "Stories" of Heaven—Clouds and Rain—The Atmospheric Circulation—Hebrew Appreciation even of the Terrible in Nature—The "Balancings" and "Spreadings" of the Clouds—The "Windows of Heaven"—Not Literal Sluice-gates—The Four Winds—The Four Quarters—The Circle of the Earth—The Waters under the Earth—The "Depths" 35

CONTENTS

Chapter V. The Ordinances of the Heavens
The Order of the Heavenly Movements—Daily Movement of the Sun—Nightly Movements of the Stars—The "Host of Heaven"—Symbolic of the Angelic Host—Morning Stars—The Scripture View of the Heavenly Order 55

Chapter VI. The Sun
The Double Purpose of the Two Great Heavenly Bodies—Symbolic Use of the Sun as Light-giver—No Deification of the Sun or of Light—Solar Idolatry in Israel—*Shemesh* and *Heres*—Sun-spots—Light before the Sun—"Under the Sun"—The Circuit of the Sun—Sunstroke—"Variableness"—Our present Knowledge of the Sun—Sir William Herschel's Theory—Conflict between the Old Science and the New—Galileo—A Question of Evidence—A Question of Principle . 63

Chapter VII. The Moon
Importance of the Moon in Olden Times—Especially to the Shepherd—Jewish Feasts at the Full Moon—The Harvest Moon—The Hebrew Month a Natural one—Different Hebrew Words for Moon—Moon-worship forbidden—"Similitudes" of the Moon—Worship of Ashtoreth—No mention of Lunar Phases—The Moon "for Seasons" 79

Chapter VIII. The Stars
Number of the Stars—"Magnitudes" of the Stars—Distances of the Stars 95

Chapter IX. Comets
Great Comets unexpected Visitors—Description of Comets—Formation of the Tail—Possible References in Scripture to Comets 103

Chapter X. Meteors
Aerolites—Diana of the Ephesians—Star-showers—The Leonid Meteors—References in Scripture—The Aurora Borealis 111

Chapter XI. Eclipses of the Sun and Moon
Vivid Impression produced by a Total Solar Eclipse—Eclipses not Omens to the Hebrews—Eclipses visible in Ancient Palestine—Explanation of Eclipses—The Saros—Scripture References to Eclipses—The Corona—The Egyptian "Winged Disc"—The Babylonian "Ring with Wings"—The Corona at Minimum . . . 118

CONTENTS

Chapter XII. Saturn and Astrology

The "Seven Planets"—Possible Scripture References to Venus and Jupiter—"Your God Remphan" probably Saturn—The Sabbath and Saturn's Day—R. A. Proctor on the Names of the Days of the Week—Order of the Planets—Alexandrian Origin of the Weekday Names—The Relation of Astrology to Astronomy—Early Babylonian Astrology—Hebrew Contempt for Divination . 130

BOOK II

THE CONSTELLATIONS

Chapter I. The Origin of the Constellations

The "Greek Sphere"—Aratus—St. Paul's Sermon at Athens—The Constellations of Ptolemy's Catalogue—References to the Constellations in Hesiod and Homer—The Constellation Figures on Greek Coins—And on Babylonian "Boundary-stones"—The Unmapped Space in the South—Its Explanation—Precession—Date and Place of the Origin of the Constellations—Significant Positions of the Serpent Forms in the Constellations—The Four "Royal Stars"—The Constellations earlier than the Old Testament 149

Chapter II. Genesis and the Constellations

The Bow set in the Cloud—The Conflict with the Serpent—The Seed of the Woman—The Cherubim—The "Mighty Hunter" 162

Chapter III. The Story of the Deluge

Resemblance between the Babylonian and Genesis Deluge Stories—The Deluge Stories in Genesis—Their Special Features—The Babylonian Deluge Story—Question as to its Date—Its Correspondence with both the Genesis Narratives—The Constellation Deluge Picture—Its Correspondence with both the Genesis Narratives—The Genesis Deluge Story independent of Star Myth and Babylonian Legend 170

Chapter IV. The Tribes of Israel and the Zodiac

Joseph's Dream—Alleged Association of the Zodiacal Figures with the Tribes of Israel—The Standards of the Four Camps of Israel—The Blessings of Jacob and Moses—The Prophecies of Balaam—The Golden Calf—The Lion of Judah 186

CONTENTS

CHAPTER V. LEVIATHAN

The Four Serpent-like Forms in the Constellations—Their Significance Positions—The Dragon's Head and Tail—The Symbols for the Nodes—The Dragon of Eclipse—Hindu Myth of Eclipses—Leviathan—References to the Stellar Serpents in Scripture—Rahab—Andromeda—"The Eyelids of the Morning"—Poetry, Science, and Myth 196

CHAPTER VI. THE PLEIADES

Difficulty of Identification—The most Attractive Constellations—*Kimah*—Not a Babylonian Star Name—A Pre-exilic Hebrew Term—The Pleiades traditionally Seven—Mädler's Suggestion—Pleiades associated in Tradition with the Rainy Season—And with the Deluge—Their "Sweet Influences"—The Return of Spring—The Pleiades in recent Photographs—Great Size and Distance of the Cluster 213

CHAPTER VII. ORION

Kesil—Probably Orion—Appearance of the Constellation—Identified in Jewish Tradition with Nimrod, who was probably Merodach—Attitude of Orion in the Sky—*Kesilim*—The "Bands" of Orion—The Bow-star and Lance-star, Orion's Dogs—Identification of Tiamat with Cetus 231

CHAPTER VIII. MAZZAROTH

Probably the "Signs of the Zodiac"—Babylonian Creation Story—Significance of its Astronomical References—Difference between the "Signs" and the "Constellations" of the Zodiac—Date of the Change—And of the Babylonian Creation Epic—Stages of Astrology—Astrology Younger than Astronomy by 2000 Years—*Mazzaroth* and the "Chambers of the South"—*Mazzaloth*—The Solar and Lunar Zodiacs—*Mazzaroth* in his Season 243

CHAPTER IX. ARCTURUS

'*Ash* and '*Ayish*—Uncertainty as to their Identification—Probably the Great Bear—*Mezarim*—Probably another Name for the Bears—"Canst thou guide the Bear?"—Proper Motions of the Plough-stars—Estimated Distance 258

CONTENTS xiii

BOOK III
TIMES AND SEASONS

CHAPTER I. THE DAY AND ITS DIVISIONS
Rotation Period of Venus—Difficulty of the Time Problem on Venus—The Sun and Stars as Time Measurers—The apparent Solar Day the First in Use—It began at Sunset—Subdivisions of the Day Interval—Between the Two Evenings—The Watches of the Night—The 12-hour Day and the 24-hour Day . . . 269

CHAPTER II. THE SABBATH AND THE WEEK
The Week not an Astronomical Period—Different Weeks employed by the Ancients—Four Origins assigned for the Week—The Quarter-month—The Babylonian System—The Babylonian Sabbath not a Rest Day—The Jewish Sabbath amongst the Romans—Alleged Astrological Origin of the Week—Origin of the Week given in the Bible 283

CHAPTER III. THE MONTH
The New Moon a Holy Day with the Hebrews—The Full Moons at the Two Equinoxes also Holy Days—The Beginnings of the Months determined from actual Observation—Rule for finding Easter—Names of the Jewish Months—Phœnician and Babylonian Month Names—Number of Days in the Month—Babylonian Dead Reckoning—Present Jewish Calendar . . 293

CHAPTER IV. THE YEAR
The Jewish Year a Luni-solar one—Need for an Intercalary Month—The Metonic Cycle—The Sidereal and Tropical Years—The Hebrew a Tropical Year—Beginning near the Spring Equinox—Meaning of "the End of the Year"—Early Babylonian Method of determining the First Month—Capella as the Indicator Star—The Triad of Stars—The Tropical Year in the Deluge Story . 305

CHAPTER V. THE SABBATIC YEAR AND THE JUBILEE
Law of the Sabbatic Year—A Year of Rest and Release—The Jubilee—Difficulties connected with the Sabbatic Year and the Jubilee—The Sabbatic Year, an Agricultural one—Interval between the Jubilees, Forty-nine Years, not Fifty—Forty-nine Years an Astronomical Cycle 326

CHAPTER VI. THE CYCLES OF DANIEL

The Jubilee Cycle possessed only by the Hebrews—High Estimation of Daniel and his Companions entertained by Nebuchadnezzar—Due possibly to Daniel's Knowledge of Luni-solar Cycles—Cycles in Daniel's Prophecy—2300 Years and 1260 Years as Astronomical Cycles—Early Astronomical Progress of the Babylonians much overrated—Yet their Real Achievements not Small—Limitations of the Babylonian—Freedom of the Hebrew   337

BOOK IV
THREE ASTRONOMICAL MARVELS

CHAPTER I. JOSHUA'S LONG DAY

METHOD OF STUDYING THE RECORD—To be discussed as it stands—An early Astronomical Observation. BEFORE THE BATTLE—Movements of the Israelites—Reasons for the Gibeonites' Action—Rapid Movements of all the Parties. DAY, HOUR, AND PLACE OF THE MIRACLE—Indication of the Sun's Declination—Joshua was at Gibeon—And at High Noon—On the 21st Day of the Fourth Month. JOSHUA'S STRATEGY—Key to it in the Flight of the Amorites by the Beth-horon Route—The Amorites defeated but not surrounded—King David as a Strategist. THE MIRACLE—The Noon-day Heat, the great Hindrance to the Israelites—Joshua desired the Heat to be tempered—The Sun made to "be silent"—The Hailstorm—The March to Makkedah—A Full Day's March in the Afternoon—"The Miracle" not a Poetic Hyperbole—Exact Accord of the Poem and the Prose Chronicle—The Record made at the Time—Their March, the Israelites' Measure of Time .   .   .   351

CHAPTER II. THE DIAL OF AHAZ

The Narrative—Suggested Explanations—The "Dial of Ahaz," probably a Staircase—Probable History and Position of the Staircase—Significance of the Sign   .   385

CHAPTER III. THE STAR OF BETHLEHEM

The Narrative—No Astronomical Details given—Purpose of the Scripture Narrative—Kepler's suggested Identification of the Star—The New Star of 1572—Legend of the Well of Bethlehem—True Significance of the Reticence of the Gospel Narrative   393

A TABLE OF SCRIPTURAL REFERENCES   .   .   .   401

INDEX   405

ILLUSTRATIONS

	PAGE
THE STAR OF BETHLEHEM (*Burne-Jones*) *Frontispiece*	
THE RAINBOW (*Rubens*) .	2
MERODACH AND TIAMAT .	25
CIRRUS AND CUMULI	47
A CORNER OF THE MILKY WAY	94
THE GREAT COMET OF 1843	102
FALL OF AN AEROLITE .	110
METEORIC SHOWER OF 1799	115
THE ASSYRIAN 'RING WITH WINGS'	126
CORONA OF MINIMUM TYPE	127
ST. PAUL PREACHING AT ATHENS (*Raphael*)	148
THE ANCIENT CONSTELLATIONS SOUTH OF THE ECLIPTIC	155
THE CELESTIAL SPHERE .	156
THE MIDNIGHT CONSTELLATIONS OF SPRING, B.C. 2700	164
THE MIDNIGHT CONSTELLATIONS OF WINTER, B.C. 2700	165
OPHIUCHUS AND THE NEIGHBOURING CONSTELLATIONS	189
AQUARIUS AND THE NEIGHBOURING CONSTELLATIONS	192
HERCULES AND DRACO .	197
HYDRA AND THE NEIGHBOURING CONSTELLATIONS	200

ILLUSTRATIONS

	PAGE
Andromeda and Cetus	207
Stars of the Pleiades	219
Inner Nebulosities of the Pleiades	227
Stars of Orion	232
Orion and the Neighbouring Constellations	236
Position of Spring Equinox, B.C. 2700	246
Position of Spring Equinox, A.D. 1900	247
Stars of the Plough, as the Winnowing Fan	263
'Blow up the Trumpet in the New Moon'	268
Position of the New Moon at the Equinoxes	316
Boundary-stone in the Louvre	318
Worship of the Sun-God at Sippara	322
'Sun, stand Thou still upon Gibeon, and Thou Moon in the Valley of Ajalon'	350
Map of Southern Palestine	357
Bearings of the Rising and Setting Points of the Sun from Gibeon	363

MERODACH AND TIAMAT. [*To face p. 2.*

Sculpture from the Palace of Assur-nazir-pel, King of Assyria. Now in the British Museum. Damaged by fire. Supposed to represent the defeat of Tiamat by Merodach.

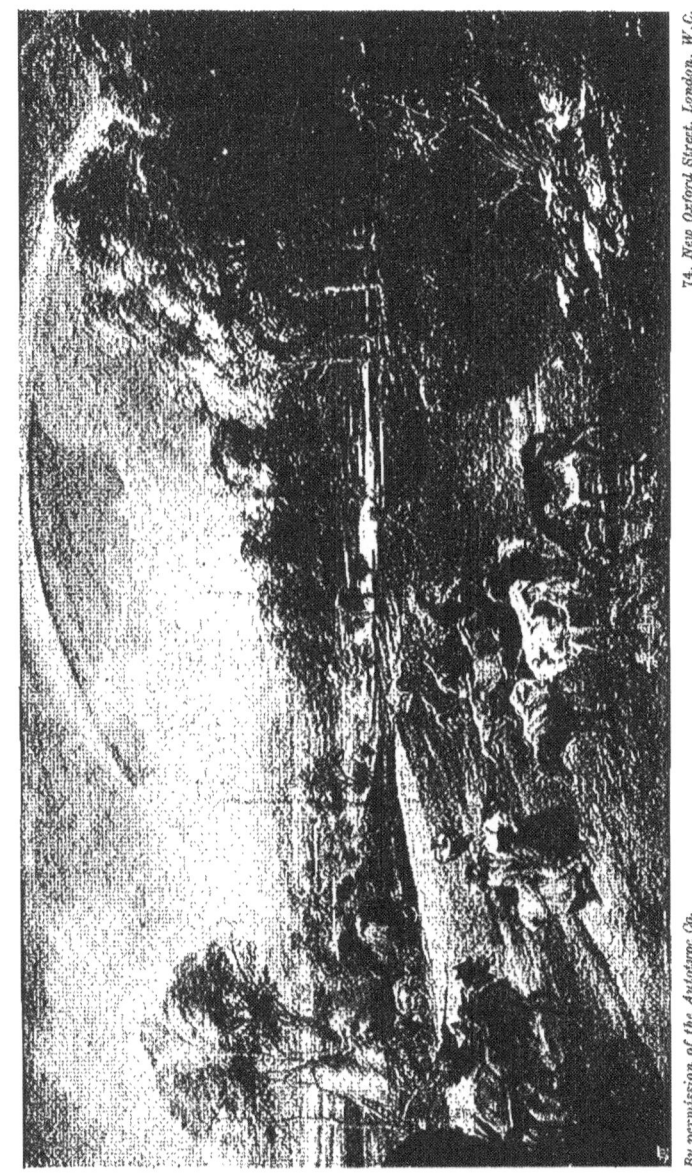

THE RAINBOW (by *Rubens*).
"The bow that is in the cloud in the day of rain."

By permission of the Autotype Co., 74, New Oxford Street, London, W.C.

THE ASTRONOMY OF THE BIBLE

BOOK I

THE HEAVENLY BODIES

CHAPTER I

THE HEBREW AND ASTRONOMY

MODERN astronomy began a little more than three centuries ago with the invention of the telescope and Galileo's application of it to the study of the heavenly bodies. This new instrument at once revealed to him the mountains on the moon, the satellites of Jupiter, and the spots on the sun, and brought the celestial bodies under observation in a way that no one had dreamed of before. In our view to-day, the planets of the solar system are worlds; we can examine their surfaces and judge wherein they resemble or differ from our earth. To the ancients they were but points of light; to us they are vast bodies that we have been able to measure and to weigh. The telescope has enabled us also to penetrate deep into outer space; we have learnt of other systems besides that of our own sun and its dependents, many of them far more complex; clusters and clouds of stars have been revealed to us, and mysterious nebulæ, which suggest by their forms that they are systems of suns in the making. More lately the invention of the spectroscope has informed us of the very elements which go to the composition of these numberless stars, and we can distinguish those which are in a similar condition to our sun from those differing from him. And photography has recorded for us objects too faint for mere sight to detect, even when aided by the most powerful telescope; too detailed and intricate for the most skilful hand to depict.

Galileo's friend and contemporary, Kepler, laid the foundations of another department of modern astronomy at about the same time. He studied the apparent movements of the planets until they yielded him their secret so far that he was able to express them in three simple laws; laws which, two generations later, Sir Isaac Newton demonstrated to be the outcome of one grand and simple law of universal range, the law of gravitation. Upon this law the marvellous mathematical conquests of astronomy have been based.

All these wonderful results have been attained by the free exercise of men's mental abilities, and it cannot be imagined that God would have intervened to hamper their growth in intellectual power by revealing to men facts and methods which it was within their own ability to discover for themselves. Men's mental powers have developed by their exercise; they would have been stunted had men been led to look to revelation rather than to diligent effort for the satisfaction of their curiosity. We therefore do not find any reference in the Bible to that which modern astronomy has taught us. Yet it may be noted that some expressions, appropriate at any time, have become much more appropriate, much more forcible, in the light of our present-day knowledge.

The age of astronomy which preceded the Modern, and may be called the Classical age, was almost as sharply defined in its beginning as its successor. It lasted about two thousand years, and began with the investigations into the movements of the planets made by some of the early Greek mathematicians. Classical, like Modern astronomy, had its two sides,—the instrumental and the mathematical. On the instrumental side was the invention of graduated instruments for the determination of the positions of the heavenly bodies; on the mathematical, the development of geometry and trigonometry for the interpretation of those positions when thus determined. Amongst the great names of this period are those of Eudoxus of Knidus (B.C. 408–355), and Hipparchus of Bithynia, who lived rather more than two centuries later. Under its first leaders astronomy in the Classical age began to advance rapidly, but it soon experienced a deadly blight. Men were not content to observe the heavenly bodies for what they were; they endeavoured to make them the sources of divination. The great school of Alexandria (founded about 300 B.C.), the headquarters of astronomy, became invaded by the spirit of astrology, the bastard science which has always tried—parasite-like—to suck its life from astronomy. Thus from the days of Claudius Ptolemy to the end of the Middle Ages the growth of astronomy was arrested, and it bore but little fruit.

It will be noticed that the Classical age did not commence until about the time of the completion of the last books of the Old Testament; so we do not find any reference in Holy Scripture to the astronomical achievements of that period, amongst which the first attempts to explain the apparent motions of sun, moon, stars, and planets were the most considerable.

We have a complete history of astronomy in the Modern and Classical periods, but there was an earlier astronomy, not inconsiderable in amount, of which no history is preserved. For when Eudoxus commenced his labours, the length of the year had already been determined, the equinoxes and solstices had been recognized, the ecliptic, the celestial equator, and the poles of both great circles were known, and the five principal planets were familiar objects. This Early astronomy must have had its history, its stages of development, but we can only with difficulty trace them out. It cannot have sprung into existence full-grown any more than the other sciences; it must have started from zero, and men must have slowly fought their way from one observation to another, with gradually widening conceptions, before they could bring it even to that stage of development in which it was when the observers of the Museum of Alexandria began their work.

The books of the Old Testament were written at different times during the progress of this Early age of astronomy. We should therefore naturally expect to find the astronomical allusions written from the standpoint of such scientific knowledge as had then been acquired. We cannot for a moment expect that any

THE HEBREW AND ASTRONOMY

supernatural revelation of purely material facts would be imparted to the writers of sacred books, two or three thousand years before the progress of science had brought those facts to light, and we ought not to be surprised if expressions are occasionally used which we should not ourselves use to-day, if we were writing about the phenomena of nature from a technical point of view. It must further be borne in mind that the astronomical references are not numerous, that they occur mostly in poetic imagery, and that Holy Scripture was not intended to give an account of the scientific achievements, if any, of the Hebrews of old. Its purpose was wholly different: it was religious, not scientific; it was meant to give spiritual, not intellectual enlightenment.

An exceedingly valuable and interesting work has recently been brought out by the most eminent of living Italian astronomers, Prof. G. V. Schiaparelli, on this subject of "Astronomy in the Old Testament," to which work I should like here to acknowledge my indebtedness. Yet I feel that the avowed object of his book,[1]—to "discover what ideas the ancient Jewish sages held regarding the structure of the universe, what observations they made of the stars, and how far they made use of them for the measurement and division of time"—is open to this criticism,—that sufficient material for carrying it out is not within our reach. If we were to accept implicitly the argument from the silence of Scripture, we should conclude that the Hebrews—though their calendar was essentially a lunar one, based upon the

[1] *Astronomy in the Old Testament*, p. 12.

actual observation of the new moon—had never noticed that the moon changed its apparent form as the month wore on, for there is no mention in the Bible of the lunar phases.

The references to the heavenly bodies in Scripture are not numerous, and deal with them either as time-measurers or as subjects for devout allusion, poetic simile, or symbolic use. But there is one characteristic of all these references to the phenomena of Nature, that may not be ignored. None of the ancients ever approached the great Hebrew writers in spiritual elevation; none equalled them in poetic sublimity; and few, if any, surpassed them in keenness of observation, or in quick sympathy with every work of the Creator.

These characteristics imply a natural fitness of the Hebrews for successful scientific work, and we should have a right to believe that under propitious circumstances they would have shown a pre-eminence in the field of physical research as striking as is the superiority of their religious conceptions over those of the surrounding nations. We cannot, of course, conceive of the average Jew as an Isaiah, any more than we can conceive of the average Englishman as a Shakespeare, yet the one man, like the other, is an index of the advancement and capacity of his race; nor could Isaiah's writings have been preserved, more than those of Shakespeare, without a true appreciation of them on the part of many of his countrymen.

But the necessary conditions for any great scientific development were lacking to Israel. A small nation, planted between powerful and aggressive empires, their history was for the most part the record of a struggle for bare existence; and after three or four centuries of the unequal conflict, first the one and then the other of the two sister kingdoms was overwhelmed. There was but little opportunity during these years of storm and stress for men to indulge in any curious searchings into the secrets of nature.

Once only was there a long interval of prosperity and peace; viz. from the time that David had consolidated the kingdom to the time when it suffered disruption under his grandson, Rehoboam; and it is significant that tradition has ascribed to Solomon and to his times just such a scientific activity as the ability and temperament of the Hebrew race would lead us to expect it to display when the conditions should be favourable for it.

Thus, in the fourth chapter of the First Book of Kings, not only are the attainments of Solomon himself described, but other men, contemporaries either of his father David or himself, are referred to, as distinguished in the same direction, though to a less degree.

"And God gave Solomon wisdom and understanding exceeding much, and largeness of heart, even as the sand that is on the seashore. And Solomon's wisdom excelled the wisdom of all the children of the east country, and all the wisdom of Egypt. For he was wiser than all men; than Ethan the Ezrahite, and Heman, and Chalcol, and Darda, the sons of Mahol: and his fame was in all nations round about. And he spake three thousand proverbs: and his songs were a thousand and five. And he spake of trees, from the cedar-tree that is in Lebanon even unto the hyssop that springeth out of the wall: he spake also of

10   THE ASTRONOMY OF THE BIBLE

beasts, and of fowl, and of creeping things, and of fishes. And there came of all people to hear the wisdom of Solomon, from all kings of the earth, which had heard of his wisdom."

The tradition of his great eminence in scientific research is also preserved in the words put into his mouth in the Book of the Wisdom of Solomon, now included in the Apocrypha.

"For" (God) "Himself gave me an unerring knowledge of the things that are, to know the constitution of the world, and the operation of the elements; the beginning and end and middle of times, the alternations of the solstices and the changes of seasons, the circuits of years and the positions" (*margin*, constellations) "of stars; the natures of living creatures and the ragings of wild beasts, the violences of winds and the thoughts of men, the diversities of plants and the virtues of roots: all things that are either secret or manifest I learned, for she that is the artificer of all things taught me, even Wisdom."

Two great names have impressed themselves upon every part of the East:—the one, that of Solomon the son of David, as the master of every secret source of knowledge; and the other that of Alexander the Great, as the mightiest of conquerors. It is not unreasonable to believe that the traditions respecting the first have been founded upon as real a basis of actual achievement as those respecting the second.

But to such scientific achievements we have no express allusion in Scripture, other than is afforded us by the two quotations just made. Natural objects, natural phenomena are not referred to for their own sake. Every

THE HEBREW AND ASTRONOMY 11

thought leads up to God, or to man's relation to Him. Nature, as a whole and in its every aspect and detail, is the handiwork of Jehovah: that is the truth which the heavens are always declaring;—and it is His power, His wisdom, and His goodness to man which it is sought to illustrate, when the beauty or wonder of natural objects is described.

> "When I consider Thy heavens, the work of Thy fingers,
> The moon and the stars, which Thou hast ordained;
> What is man, that Thou art mindful of him?
> And the son of man, that Thou visitest him?"

The first purpose, therefore, of the following study of the astronomy of the Bible is,—not to reconstruct the astronomy of the Hebrews, a task for which the material is manifestly incomplete,—but to examine such astronomical allusions as occur with respect to their appropriateness to the lesson which the writer desired to teach. Following this, it will be of interest to examine what connection can be traced between the Old Testament Scriptures and the Constellations; the arrangement of the stars into constellations having been the chief astronomical work effected during the centuries when those Scriptures were severally composed. The use made of the heavenly bodies as time-measurers amongst the Hebrews will form a third division of the subject; whilst there are two or three incidents in the history of Israel which appear to call for examination from an astronomical point of view, and may suitably be treated in a fourth and concluding section.

CHAPTER II

THE CREATION

A FEW years ago a great eclipse of the sun, seen as total along a broad belt of country right across India, drew thither astronomers from the very ends of the earth. Not only did many English observers travel thither, but the United States of America in the far west, and Japan in the far east sent their contingents, and the entire length of country covered by the path of the shadow was dotted with the temporary observatories set up by the men of science.

It was a wonderful sight that was vouchsafed to these travellers in pursuit of knowledge. In a sky of unbroken purity, undimmed even for a moment by haze or cloud, there shone down the fierce Indian sun. Gradually a dark mysterious circle invaded its lower edge, and covered its brightness; coolness replaced the burning heat; slowly the dark covering crept on; slowly the sunlight diminished until at length the whole of the sun's disc was hidden. Then in a moment a wonderful starlike form flashed out, a noble form of glowing silver light on the deep purple-coloured sky.

There was, however, no time for the astronomers to devote

to admiration of the beauty of the scene, or indulgence in rhapsodies. Two short minutes alone were allotted them to note all that was happening, to take all their photographs, to ask all the questions, and obtain all the answers for which this strange veiling of the sun, and still stranger unveiling of his halo-like surroundings, gave opportunity. It was two minutes of intensest strain, of hurried though orderly work; and then a sudden rush of sunlight put an end to all. The mysterious vision had withdrawn itself; the colour rushed back to the landscape, so corpse-like whilst in the shadow; the black veil slid rapidly from off the sun; the heat returned to the air; the eclipse was over.

But the astronomers from distant lands were not the only people engaged in watching the eclipse. At their work, they could hear the sound of a great multitude, a sound of weeping and wailing, a people dismayed at the distress of their god.

It was so at every point along the shadow track, but especially where that track met the course of the sacred river. Along a hundred roads the pilgrims had poured in unceasing streams towards Holy Mother Gunga; towards Benares, the sacred city; towards Buxar, where the eclipse was central at the river bank. It is always meritorious—so the Hindoo holds—to bathe in that sacred river, but such a time as this, when the sun is in eclipse, is the most propitious moment of all for such lustration.

Could there be a greater contrast than that offered between the millions trembling and dismayed at the signs of heaven, and the little companies who had come for thousands of miles over land and sea, rejoicing in the brief chance that was given them for learning a little more of the secrets of the wonders of Nature?

The contrast between the heathen and the scientists was in both their spiritual and their intellectual standpoint, and, as we shall see later, the intellectual contrast is a result of the spiritual. The heathen idea is that the orbs of heaven are divine, or at least that each expresses a divinity. This does not in itself seem an unnatural idea when we consider the great benefits that come to us through the instrumentality of the sun and moon. It is the sun that morning by morning rolls back the darkness, and brings light and warmth and returning life to men; it is the sun that rouses the earth after her winter sleep and quickens vegetation. It is the moon that has power over the great world of waters, whose pulse beats in some kind of mysterious obedience to her will.

Natural, then, has it been for men to go further, and to suppose that not only is power lodged in these, and in the other members of the heavenly host, but that it is living, intelligent, personal power; that these shining orbs are beings, or the manifestations of beings; exalted, mighty, immortal;—that they are gods.

But if these are gods, then it is sacrilegious, it is profane, to treat them as mere "things"; to observe them minutely in the microscope or telescope; to dissect them, as it were, in the spectroscope; to identify their elements in the laboratory; to be curious about their properties, influences, relations, and actions on each other.

THE CREATION

And if these are gods, there are many gods, not One God. And if there are many gods, there are many laws, not one law. Thus scientific observations cannot be reconciled with polytheism, for scientific observations demand the assumption of one universal law. The wise king expressed this law thus :—

"The thing that hath been, it is that which shall be." The actual language of science, as expressed by Professor Thiele, a leading Continental astronomer, states that—

"Everything that exists, and everything that happens, exists or happens as a necessary consequence of a previous state of things. If a state of things is repeated in every detail, it must lead to exactly the same consequences. Any difference between the results of causes that are in part the same, must be explainable by some difference in the other part of the causes."[1]

The law stated in the above words has been called the Law of Causality. It "cannot be proved, but must be believed; in the same way as we believe the fundamental assumptions of religion, with which it is closely and intimately connected. The law of causality forces itself upon our belief. It may be denied in theory, but not in practice. Any person who denies it, will, if he is watchful enough, catch himself constantly asking himself, if no one else, why *this* has happened, and not *that*. But in that very question he bears witness to the law of causality. If we are consistently to deny the law of causality, we must repudiate all observation, and particularly all prediction based on past experience, as useless and misleading.

"If we could imagine for an instant that the same complete combination of causes could have a definite number of different consequences, however small that

---

[1] T. N. Thiele, Director of the Copenhagen Observatory, *Theory of Observations*, p. 1.

16 THE ASTRONOMY OF THE BIBLE

number might be, and that among these the occurrence of the actual consequence was, in the old sense of the word, accidental, no observation would ever be of any particular value." [1]

So long as men hold, as a practical faith, that the results which attend their efforts depend upon whether Jupiter is awake and active, or Neptune is taking an unfair advantage of his brother's sleep; upon whether Diana is bending her silver bow for the battle, or flying weeping and discomfited because Juno has boxed her ears—so long is it useless for them to make or consult observations.

But, as Professor Thiele goes on to say—

"If the law of causality is acknowledged to be an assumption which always holds good, then every observation gives us a revelation which, when correctly appraised and compared with others, teaches us the laws by which God rules the world."

By what means have the modern scientists arrived at a position so different from that of the heathen? It cannot have been by any process of natural evolution that the intellectual standpoint which has made scientific observation possible should be derived from the spiritual standpoint of polytheism which rendered all scientific observation not only profane but useless.

In the old days the heathen in general regarded the heavenly host and the heavenly bodies as the heathen do

[1] T. N. Thiele, Director of the Copenhagen Observatory, *Theory of Observations*, p. 1.

to-day. But by one nation, the Hebrews, the truth that—

"In the beginning God created the heaven and the earth"

was preserved in the first words of their Sacred Book. That nation declared—

"All the gods of the people are idols: but the Lord made the heavens."

For that same nation the watchword was—

"Hear, O Israel: the Lord our God is one Lord."

From these words the Hebrews not only learned a great spiritual truth, but derived intellectual freedom. For by these words they were taught that all the host of heaven and of earth were created things—merely "things," not divinities—and not only that, but that the Creator was One God, not many gods; that there was but one law-giver; and that therefore there could be no conflict of laws. These first words of Genesis, then, may be called the charter of all the physical sciences, for by them is conferred freedom from all the bonds of unscientific superstition, and by them also do men know that consistent law holds throughout the whole universe. It is the intellectual freedom of the Hebrew that the scientist of to-day inherits. He may not indeed be able to rise to the spiritual standpoint of the Hebrew, and consciously acknowledge that—

"Thou, even Thou, art Lord alone; Thou hast made heaven, the heaven of heavens, with all their host, the earth, and all things that are therein, the seas, and all that is therein, and Thou preservest them all; and the host of heaven worshippeth Thee."

But he must at least unconsciously assent to it, for it is on the first great fundamental assumption of religion as stated in the first words of Genesis, that the fundamental assumption of all his scientific reasoning depends.

Scientific reasoning and scientific observation can only hold good so long and in so far as the Law of Causality holds good. We must assume a pre-existing state of affairs which has given rise to the observed effect; we must assume that this observed effect is itself antecedent to a subsequent state of affairs. Science therefore cannot go back to the absolute beginnings of things, or forward to the absolute ends of things. It cannot reason about the way matter and energy came into existence, or how they might cease to exist; it cannot reason about time or space, as such, but only in the relations of these to phenomena that can be observed. It does not deal with things themselves, but only with the relations between things. Science indeed can only consider the universe as a great machine which is in "going order," and it concerns itself with the relations which some parts of the machine bear to other parts, and with the laws and manner of the "going" of the machine in those parts. The relations of the various parts, one to the other, and the way in which they work together, may afford some idea of the design and purpose of the machine, but it can give no information as to how the material of which it is composed came into existence, nor as to the method by which it was originally constructed. Once started, the machine comes under the scrutiny of science, but the actual starting lies outside its scope.

Men therefore cannot find out for themselves how the worlds were originally made, how the worlds were first moved, or how the spirit of man was first formed within him; and this, not merely because these beginnings of things were of necessity outside his experience, but also because beginnings, as such, must lie outside the law by which he reasons.

By no process of research, therefore, could man find out for himself the facts that are stated in the first chapter of Genesis. They must have been revealed. Science cannot inquire into them for the purpose of checking their accuracy; it must accept them, as it accepts the fundamental law that governs its own working, without the possibility of proof.

And this is what has been revealed to man:—that the heaven and the earth were not self-existent from all eternity, but were in their first beginning created by God. As the writer of the Epistle to the Hebrews expresses it: "Through faith we understand that the worlds were framed by the word of God, so that things which are seen were not made of things which do appear." And a further fact was revealed that man could not have found out for himself; viz. that this creation was made and finished in six Divine actings, comprised in what the narrative denominates "days." It has not been revealed whether the duration of these "days" can be expressed in any astronomical units of time.

Since under these conditions science can afford no information, it is not to be wondered at that the hypotheses that have been framed from time to time to "explain" the first chapter of Genesis, or to express it in scientific terms, are not wholly satisfactory. At one time the chapter was interpreted to mean that the entire universe was called into existence about 6,000 years ago, in six days of twenty-four hours each. Later it was recognized that both geology and astronomy seemed to indicate the existence of matter for untold millions of years instead of some six thousand. It was then pointed out that, so far as the narrative was concerned, there might have been a period of almost unlimited duration between its first verse and its fourth; and it was suggested that the six days of creation were six days of twenty-four hours each, in which, after some great cataclysm, 6,000 years ago, the face of the earth was renewed and replenished for the habitation of man, the preceding geological ages being left entirely unnoticed. Some writers have confined the cataclysm and renewal to a small portion of the earth's surface—to "Eden," and its neighbourhood. Other commentators have laid stress on the truth revealed in Scripture that "one day is with the Lord as a thousand years, and a thousand years as one day," and have urged the argument that the six days of creation were really vast periods of time, during which the earth's geological changes and the evolution of its varied forms of life were running their course. Others, again, have urged that the six days of creation were six literal days, but instead of being consecutive were separated by long ages. And yet again, as no man was present during the creation period, it has been suggested that the Divine revelation of it was given to Moses or some other inspired prophet in six successive visions or dreams, which constituted the "six days" in which the chief facts of creation were set forth.

All such hypotheses are based on the assumption that the opening chapters of Genesis are intended to reveal to man certain physical details in the material history of this planet; to be in fact a little compendium of the geological and zoological history of the world, and so a suitable introduction to the history of the early days of mankind which followed it.

It is surely more reasonable to conclude that there was no purpose whatever of teaching us anything about the physical relationships of land and sea, of tree and plant, of bird and fish; it seems, indeed, scarcely conceivable that it should have been the Divine intention so to supply the ages with a condensed manual of the physical sciences. What useful purpose could it have served? What man would have been the wiser or better for it? Who could have understood it until the time when men, by their own intellectual strivings, had attained sufficient knowledge of their physical surroundings to do without such a revelation at all?

But although the opening chapters of Genesis were not designed to teach the Hebrew certain physical facts of nature, they gave him the knowledge that he might lawfully study nature. For he learnt from them that nature has no power nor vitality of its own; that sun, and sea, and cloud, and wind are not separate deities, nor the expression of deities; that they are but "things," however glorious and admirable; that they are the handiwork of God; and—

"The works of the Lord are great,
    Sought out of all them that have pleasure therein.
His work is honour and majesty;
    And His righteousness endureth for ever.
He hath made His wonderful works to be remembered."

What, then, is the significance of the detailed account given us of the works effected on the successive days of creation? Why are we told that light was made on the first day, the firmament on the second, dry land on the third, and so on? Probably for two reasons. First, that the rehearsal, as in a catalogue, of the leading classes of natural objects, might give definiteness and precision to the teaching that each and all were creatures, things made by the word of God. The bald statement that the heaven and the earth were made by God might still have left room for the imagination that the powers of nature were co-eternal with God, or were at least subordinate divinities; or that other powers than God had worked up into the present order the materials He had created. The detailed account makes it clear that not only was the universe in general created by God, but that there was no part of it that was not fashioned by Him.

The next purpose was to set a seal of sanctity upon the Sabbath. In the second chapter of Genesis we read—

"On the seventh day God ended His work which He had made; and He rested on the seventh day from all

THE CREATION

His work which He had made. And God blessed the seventh day, and sanctified it: because that in it He had rested from all His work which God created and made."

In this we get the institution of the *week*, the first ordinance imposed by God upon man. For in the fourth of the ten commandments which God gave through Moses, it is said—

"The seventh day is the sabbath of the Lord thy God: in it thou shalt not do any work. . . . For in six days the Lord made heaven and earth, the sea, and all that in them is, and rested the seventh day: wherefore the Lord blessed the sabbath day, and hallowed it."

And again, when the tabernacle was being builded, it was commanded—

"The children of Israel shall keep the sabbath, to observe the sabbath throughout their generations, for a perpetual covenant. It is a sign between Me and the children of Israel for ever: for in six days the Lord made heaven and earth, and on the seventh day He rested, and was refreshed."

God made the sun, moon, and stars, and appointed them "for signs, and for seasons, and for days, and years." The sun marks out the days; the moon by her changes makes the months; the sun and the stars mark out the seasons and the years. These were divisions of time which man would naturally adopt. But there is not an exact number of days in the month, nor an exact number of days or months in the year. Still less does the period of seven days fit precisely into month or season or year; the week is marked out by no phase of the moon, by no fixed relation between the sun, the moon, or the stars. It is not a division of time that man would naturally adopt for himself; it runs across all the natural divisions of time.

What are the six days of creative work, and the seventh day—the Sabbath—of creative rest? They are not days of man, they are days of God; and our days of work and rest, our week with its Sabbath, can only be the figure and shadow of that week of God; something by which we may gain some faint apprehension of its realities, not that by which we can comprehend and measure it.

Our week, therefore, is God's own direct appointment to us; and His revelation that He fulfilled the work of creation in six acts or stages, dignifies and exalts the toil of the labouring man, with his six days of effort and one of rest, into an emblem of the creative work of God.

CHAPTER III

THE DEEP

THE second verse of Genesis states, "And the earth was without form and void [*i.e.* waste and empty] and darkness was upon the face of the deep." The word *tehōm*, here translated *deep*, has been used to support the theory that the Hebrews derived their Creation story from one which, when exiles in Babylon, they heard from their conquerors. If this theory were substantiated, it would have such an important bearing upon the subject of the attitude of the inspired writers towards the objects of nature, that a little space must be spared for its examination.

The purpose of the first chapter of Genesis is to tell us that—

"In the beginning God created the heaven and the earth."

From it we learn that the universe and all the parts that make it up—all the different forms of energy, all the different forms of matter—are neither deities themselves, nor their embodiments and expressions, nor the work of conflicting deities. From it we learn that the universe is not self-existent, nor even (as the pantheist thinks of it) the expression of one vague, impersonal and unconscious, but all-pervading influence. It was not self-made; it did not exist from all eternity. It is not God, for God made it.

But the problem of its origin has exercised the minds of many nations beside the Hebrews, and an especial interest attaches to the solution arrived at by those nations who were near neighbours of the Hebrews and came of the same great Semitic stock.

From the nature of the case, accounts of the origin of the world cannot proceed from experience, or be the result of scientific experiment. They cannot form items of history, or arise from tradition. There are only two possible sources for them; one, Divine revelation; the other, the invention of men.

The account current amongst the Babylonians has been preserved to us by the Syrian writer Damascius, who gives it as follows:—

"But the Babylonians, like the rest of the Barbarians, pass over in silence the one principle of the Universe, and they constitute two, Tavthê and Apasôn, making Apasôn the husband of Tavthê, and denominating her " the mother of the gods." And from these proceeds an only-begotten son, Mumis, which, I conceive, is no other than the intelligible world proceeding from the two principles. From them also another progeny is derived, Lakhê and Lakhos; and again a third, Kissarê and Assôros, from which last three others proceed, Anos and Illinos and Aos. And of Aos and Dakhê is born a son called Bêlos, who, they say, is the fabricator of the world."[1]

[1] *Records of the Past*, vol. i. p. 124.

The actual story, thus summarized by Damascius, was discovered by Mr. George Smith, in the form of a long epic poem, on a series of tablets, brought from the royal library of Kouyunjik, or Nineveh, and he published them in 1875, in his book on *The Chaldean Account of Genesis*. None of the tablets were perfect; and of some only very small portions remain. But portions of other copies of the poem have been discovered in other localities, and it has been found possible to piece together satisfactorily a considerable section, so that a fair idea of the general scope of the poem has been given to us.

It opens with the introduction of a being, Tiamtu—the Tavthê of the account of Damascius,—who is regarded as the primeval mother of all things.

"When on high the heavens were unnamed,
Beneath the earth bore not a name:
The primeval ocean was their producer;
Mummu Tiamtu was she who begot the whole of them.
Their waters in one united themselves, and
The plains were not outlined, marshes were not to be seen.
When none of the gods had come forth,
They bore no name, the fates (had not been determined)
There were produced the gods (all of them)."[1]

The genealogy of the gods follows, and after a gap in the story, Tiamat, or Tiamtu, is represented as preparing for battle, "She who created everything . . . produced giant serpents." She chose one of the gods, Kingu, to be her husband and the general of her forces, and delivered to him the tablets of fate.

The second tablet shows the god Anšar, angered at the

[1] *The Old Testament in the Light of the Historical Records of Assyria and Babylonia*, by T. G. Pinches, p. 16.

threatening attitude of Tiamat, and sending his son Anu to speak soothingly to her and calm her rage. But first Anu and then another god turned back baffled, and finally Merodach, the son of Ea, was asked to become the champion of the gods. Merodach gladly consented, but made good terms for himself. The gods were to assist him in every possible way by entrusting all their powers to him, and were to acknowledge him as first and chief of all. The gods in their extremity were nothing loth. They feasted Merodach, and, when swollen with wine, endued him with all magical powers, and hailed him—

"Merodach, thou art he who is our avenger,
(Over) the whole universe have we given thee the kingdom." [1]

At first the sight of his terrible enemy caused even Merodach to falter, but plucking up courage he advanced to meet her, caught her in his net, and, forcing an evil wind into her open mouth—

"He made the evil wind enter so that she could not close her lips.
The violence of the winds tortured her stomach, and
her heart was prostrated and her mouth was twisted.
He swung the club, he shattered her stomach;
he cut out her entrails; he over-mastered (her) heart;
he bound her and ended her life.
He threw down her corpse; he stood upon it." [2]

The battle over and the enemy slain, Merodach considered how to dispose of the corpse.

"He strengthens his mind, he forms a clever plan,
And he stripped her of her skin like a fish, according to his plan." [3]

---

[1] *The Old Testament in the Light of the Historical Records of Assyria and Babylonia*, by T. G. Pinches, p. 16.
[2] *Records of the Past*, vol. i. p. 140.   [3] *Ibid.* p. 142.

Of one half of the corpse of Tiamat he formed the earth, and of the other half the heavens. He then proceded to furnish the heavens and the earth with their respective equipments; the details of this work occupying apparently the fifth, sixth, and seventh tablets of the series.

Under ordinary circumstances such a legend as the foregoing would not have attracted much attention. It is as barbarous and unintelligent as any myth of Zulu or Fijian. Strictly speaking, it is not a Creation myth at all Tiamat and her serpent-brood and the gods are all existent before Merodach commences his work, and all that the god effects is a reconstruction of the world. The method of this reconstruction possesses no features superior to those of the Creation myths of other barbarous nations. Our own Scandinavian ancestors had a similar one, the setting of which was certainly not inferior to the grotesque battle of Merodach with Tiamat. The prose Edda tells us that the first man, Bur, was the father of Bör, who was in turn the father of Odin and his two brothers Vili and Ve. These sons of Bör slew Ymir, the old frost giant.

"They dragged the body of Ymir into the middle of Ginnungagap, and of it formed the earth. From Ymir's blood they made the sea and waters; from his flesh, the land; from his bones, the mountains; and his teeth and jaws, together with some bits of broken bones, served them to make the stones and pebbles."

It will be seen that there is a remarkable likeness between the Babylonian and Scandinavian myths in the central and essential feature of each, viz. the way in which the world is supposed to have been built up by the gods from the fragments of the anatomy of a huge primæval monster. Yet it is not urged that there is any direct genetic connection between the two; that the Babylonians either taught their legend to the Scandinavians or learnt it from them.

Under ordinary circumstances it would hardly have occurred to any one to try to derive the monotheistic narrative of Gen. i. from either of these pagan myths, crowded as they are with uncouth and barbarous details. But it happened that Mr. George Smith, who brought to light the Assyrian Creation tablets, brought also to light a Babylonian account of the Flood, which had a large number of features in common with the narrative of Gen. vi.–ix. The actual resemblance between the two Deluge narratives has caused a resemblance to be imagined between the two Creation narratives. It has been well brought out in some of the later comments of Assyriologists that, so far from there being any resemblance in the Babylonian legend to the narrative in Genesis, the two accounts differ *in toto*. Mr. T. G. Pinches, for example, points out that in the Babylonian account there is—

"No direct statement of the creation of the heavens and the earth;
"No systematic division of the things created into groups and classes, such as is found in Genesis;
"No reference to the Days of Creation.;
"No appearance of the Deity as the first and only cause of the existence of things."[1]

[1] *The Old Testament in the Light of the Historical Records of Assyria and Babylonia*, by T. G. Pinches, p. 49.

Indeed, in the Babylonian account, "the heavens and the earth are represented as existing, though in a chaotic form, from the first."

Yet on this purely imaginary resemblance between the Biblical and Babylonian Creation narratives the legend has been founded "that the introductory chapters of the Book of Genesis present to us the Hebrew version of a mythology common to many of the Semitic peoples." And the legend has been yet further developed, until writers of the standing of Prof. Friedrich Delitzsch have claimed that the Genesis narrative was *borrowed* from the Babylonian, though "the priestly scholar who composed Genesis, chapter i. endeavoured of course to remove all possible mythological features of this Creation story." [1]

If the Hebrew priest did borrow from the Babylonian myth, what was it that he borrowed? Not the existence of sea and land, of sun and moon, of plants and animals, of birds and beasts and fishes. For surely the Hebrew may be credited with knowing this much of himself, without any need for a transportation to Babylon to learn it. "In writing an account of the Creation, statements as to what are the things created must of necessity be inserted," [2] whenever, wherever, and by whomsoever that account is written.

What else, then, is there common to the two accounts? *Tiamat* is the name given to the Babylonian mother of the universe, the dragon of the deep; and in Genesis

---

[1] *Babel and Bible*, Johns' translation, pp. 36 and 37.
[2] *The Old Testament in the Light of the Historical Records of Assyria and Babylonia*, by T. G. Pinches, p. 48.

it is written that "darkness was upon the face of the *deep* (*tehōm*)."

Here, and here only, is a point of possible connection; but if it be evidence of a connection, what kind of a connection does it imply? It implies that the Babylonian based his barbarous myth upon the Hebrew narrative. There is no other possible way of interpreting the connection,—if connection there be.

The Hebrew word would seem to mean, etymologically, "*surges*," "*storm-tossed waters*,"—"Deep calleth unto deep at the noise of Thy waterspouts." Our word "*deep*" is apt to give us the idea of stillness—we have the proverb, "Still waters run deep,"—whereas in some instances *tehōm* is used in Scripture of waters which were certainly shallow, as, for instance, those passed through by Israel at the Red Sea :—

"Pharaoh's chariots and his host hath He cast into the sea: his chosen captains also are drowned in the Red Sea. The *depths* have covered them."

In other passages the words used in our Authorized Version, "*deep*" or "*depths*," give the correct signification.

But deep waters, or waters in commotion, are in either case natural objects. We get the word *tehōm* used continually in Scripture in a perfectly matter-of-fact way, where there is no possibility of personification or myth being intended. Tiamat, on the contrary, the Babylonian dragon of the waters, is a mythological personification. Now the natural object must come first. It never yet has been the case that a nation has gained its knowledge of a perfectly common natural object by de-mythologizing one of the mythological personifications of another nation. The Israelites did not learn about *tehōm*, the surging water of the Red Sea, that rolled over the Egyptians in their sight, from any Babylonian fable of a dragon of the waters, read by their descendants hundreds of years later.

Yet further, the Babylonian account of Creation is comparatively late; the Hebrew account, as certainly, comparatively early. It is not merely that the actual cuneiform tablets are of date about 700 B.C., coming as they do from the Kouyunjik mound, the ruins of the palace of Sennacherib and Assurbanipal, built about that date. The poem itself, as Prof. Sayce has pointed out, indicates, by the peculiar pre-eminence given in it to Merodach, that it is of late composition. It was late in the history of Babylon that Merodach was adopted as the supreme deity. The astronomical references in the poem are more conclusive still, for, as will be shown later on, they point to a development of astronomy that cannot be dated earlier than 700 B.C.

On the other hand, the first chapter of Genesis was composed very early. The references to the heavenly bodies in verse 16 bear the marks of the most primitive condition possible of astronomy. The heavenly bodies are simply the greater light, the lesser light, and the stars —the last being introduced quite parenthetically. It is the simplest reference to the heavenly bodies that is made in Scripture, or that, indeed, could be made.

There may well have been Babylonians who held higher

conceptions of God and nature than those given in the Tiamat myth. It is certain that very many Hebrews fell short of the teaching conveyed in the first chapter of Genesis. But the fact remains that the one nation preserved the Tiamat myth, the other the narrative of Genesis, and each counted its own Creation story sacred. We can only rightly judge the two nations by what they valued. Thus judged, the Hebrew nation stands as high above the Babylonian in intelligence, as well as in faith, as the first chapter of Genesis is above the Tiamat myth.

CHAPTER IV

THE FIRMAMENT

THE sixth verse of the first chapter of Genesis presents a difficulty as to the precise meaning of the principal word, viz. that translated *firmament*.

"And God said, Let there be a *rāqiā'* in the midst of the waters, and let it divide the waters from the waters. And God made the *rāqiā'*, and divided the waters which were under the *rāqiā'* from the waters which were above the *rāqiā'*: and it was so. And God called the *rāqiā' Shamayim*. And the evening and the morning were the second day."

It is, of course, perfectly clear that by the word *rāqiā'* in the preceding passage it is the atmosphere that is alluded to. But later on in the chapter the word is used in a slightly different connection. "God said, Let there be lights in the firmament of heaven."

As we look upward from the earth, we look through a twofold medium. Near the earth we have our atmosphere; above that there is inter-stellar space, void of anything, so far as we know, except the Ether. We are not able to detect any line of demarcation where our atmosphere ends, and the outer void begins. Both therefore are equally spoken of as "the firmament"; and yet there is a difference between the two. The lower supports the clouds; in the upper are set the two great lights and the stars. The upper, therefore, is emphatically *reqiā' hasshamayim*, "the firmament of heaven," of the "uplifted." It is "in the face of"—that is, "before," or "under the eyes of," "beneath,"—this higher expanse that the fowls of the air fly to and fro.

The firmament, then, is that which Tennyson sings of as "the central blue," the seeming vault of the sky, which we can consider as at any height above us that we please. The clouds are above it in one sense; yet in another, sun, moon and stars, which are clearly far higher than the clouds, are set in it.

There is no question therefore as to what is referred to by the word "firmament"; but there is a question as to the etymological meaning of the word, and associated with that, a question as to how the Hebrews themselves conceived of the celestial vault.

The word *rāqiā'*, translated "firmament," properly signifies "an expanse," or "extension," something stretched or beaten out. The verb from which this noun is derived is often used in Scripture, both as referring to the heavens and in other connections. Thus in Job xxxvii. 18, the question is asked, "Canst thou with Him *spread out* the sky, which is strong as a molten mirror?" Eleazar, the priest, after the rebellion of Korah, Dathan and Abiram took the brazen censers of the rebels, and they were "*made broad* plates for a covering of the altar." The goldsmith described by Isaiah as making an idol, "*spreadeth it over* with gold"; whilst Jeremiah says,

THE FIRMAMENT

"silver *spread* into plates is brought from Tarshish." Again, in Psalm cxxxvi., in the account of creation we have the same word used with reference to the earth, "To Him that *stretched out* the earth above the waters." In this and in many other passages the idea of extension is clearly that which the word is intended to convey. But the Seventy, in making the Greek Version of the Old Testament, were naturally influenced by the views of astronomical science then held in Alexandria, the centre of Greek astronomy. Here, and at this time, the doctrine of the crystalline spheres—a misunderstanding of the mathematical researches of Eudoxus and others—held currency. These spheres were supposed to be a succession of perfectly transparent and invisible solid shells, in which the sun, moon, and planets were severally placed. The Seventy no doubt considered that in rendering *rāqiā'*, by *stereōma*, i. e. firmament, thus conveying the idea of a solid structure, they were speaking the last word of up-to-date science.

There should be no reluctance in ascribing to the Hebrews an erroneous scientific conception if there is any evidence that they held it. We cannot too clearly realize that the writers of the Scriptures were not supernaturally inspired to give correct technical scientific descriptions; and supposing they had been so inspired, we must bear in mind that we should often consider those descriptions wrong just in proportion to their correctness, for the very sufficient reason that not even our own science of to-day has yet reached finality in all things.

There should be no reluctance in ascribing to the

Hebrews an erroneous scientific conception if there is any evidence that they held it. In this case, there is no such evidence; indeed, there is strong evidence to the contrary.

The Hebrew word *rāqīāʻ*, as already shown, really signifies "extension," just as the word for heaven, *shamayim* means the "uplifted." In these two words, therefore, significant respectively of a surface and of height, there is a recognition of the "three dimensions,"—in other words, of Space.

When we wish to refer to super-terrestrial space, we have two expressions in modern English by which to describe it: we can speak of "the vault of heaven," or of "the canopy of heaven." "The vault of heaven" is most used, it has indeed been recently adopted as the title of a scientific work by a well-known astronomer. But the word *vault* certainly gives the suggestion of a solid structure; whilst the word *canopy* calls up the idea of a slighter covering, probably of some textile fabric.

The reasons for thinking that the Hebrews did not consider the "firmament" a solid structure are, first, that the word does not necessarily convey that meaning; next, that the attitude of the Hebrew mind towards nature was not such as to require this idea. The question, "What holds up the waters above the firmament?" would not have troubled them. It would have been sufficient for them, as for the writer to the Hebrews, to consider that God was "upholding all things by the word of His power," and they would not have troubled about the machinery. But besides this, there are many passages in Scripture, some occurring in the earliest books, which expressly speak of the clouds as carrying the water; so that the expressions placing waters "above the firmament," or "above the heavens," can mean no more than "in the clouds." Indeed, as we shall see, quite a clear account is given of the atmospheric circulation, such as could hardly be mended by a modern poet.

It is true that David sang that "the *foundations* of heaven moved and shook, because He was wroth," and Job says, that "the *pillars* of heaven tremble and are astonished at His reproof." But not only are the references to foundations and pillars evidently intended merely as poetic imagery, but they are also used much more frequently of the earth, and yet at the same time Job expressly points out that God "stretcheth out the north over the empty place, and hangeth the earth upon nothing." The Hebrew formed no ideas like those of the Hindus, who thought the earth supported by elephants, the elephants by a tortoise, the tortoise by a snake.

In Scripture, in most cases the word "earth" (*eretz*) does not mean the solid mass of this our planet, but only its surface; the "dry land" as opposed to the "seas"; the countries, the dwelling place of man and beast. The "pillars" or "foundations" of the earth in this sense are the great systems of the rocks, and these were conceived of as directly supported by the power of God, without any need of intermediary structures. The Hebrew clearly recognized that it is the will of God alone that keeps the whole secure.

Thus Hannah sang—

> "The pillars of the earth are the Lord's,
> And He hath set the world upon them."

And Asaph represents the Lord as saying:—
"The earth and all the inhabitants thereof are dissolved :
  I bear up the pillars of it."

Yet again, just as we speak of "the celestial canopy," so Psalm civ. describes the Lord as He "who stretchest out the heavens like a curtain," and Isaiah gives the image in a fuller form,—"that stretcheth out the heavens as a curtain, and spreadeth them out as a tent to dwell in." The same expression of "stretching out the heavens" is repeatedly used in Isaiah; it is indeed one of his typical phrases. Here, beyond question, extension, spreading out, is the idea sought to be conveyed, not that of solidity.

The prophet Amos uses yet another parallel. "It is He that buildeth His stories in the heaven." While Isaiah speaks of the entire stellar universe as the tent or pavilion of Jehovah, Amos likens the height of the heavens as the steps up to His throne; the "stories" are the "ascent," as Moses speaks of the "ascent of Akrabbim," and David makes "the ascent" of the Mount of Olives. The Hebrews cannot have regarded the heavens as, literally, both staircase and reservoir.

The firmament, *i.e.* the atmosphere, is spoken of as dividing between the waters that are under the firmament, *i.e.* oceans, seas, rivers, etc., from the waters that are above the firmament, *i.e.* the masses of water vapour carried by the atmosphere, seen in the clouds, and condensing from them as rain. We get the very same expression as this of the " waters which were above " in the Psalm of Praise :—

THE FIRMAMENT 41

> "Praise Him, ye heavens of heavens,
> And ye waters that be above the heavens;"

and again in the Song of the Three Children:—

"O all ye waters that be above the heaven, bless ye the Lord."

In the later books of the Bible the subject of the circulation of water through the atmosphere is referred to much more fully. Twice over the prophet Amos describes Jehovah as "He that calleth for the waters of the sea, and poureth them out upon the face of the earth." This is not merely a reference to the tides, for the Preacher in the book of Ecclesiastes expressly points out that "all the rivers run into the sea; yet the sea is not full; unto the place from whence the rivers come, thither they return again"; and Isaiah seems to employ something of the same thought:

"For as the rain cometh down and the snow from heaven, and returneth not thither, but watereth the earth, and maketh it bring forth and bud, and giveth seed to the sower and bread to the eater."

Schiaparelli indeed argues that this very passage from Isaiah "expressly excludes any idea of an atmospheric circulation of waters"[1] on the ground that the water so falling is thought to be transmuted into seeds and fruits. But surely the image is as true as it is beautiful! The rain is absorbed by vegetation, and is transmuted into seeds and fruit, and it would go hard to say that the same particles of rain are again evaporated and

---
[1] *Astronomy in the Old Testament*, p. 33 note.

taken up afresh into the clouds. Besides, if we complete the quotation we find that what is stated is that the rain does not return *until* it has accomplished its purpose :—

"So shall My word be that goeth forth out of My mouth : it shall not return unto Me void, but it shall accomplish that which I please, and it shall prosper in the thing whereto I sent it."

Elihu describes the process of evaporation precisely :—

"Behold, God is great, and we know Him not ;
The number of His years is unsearchable.
For He draweth up the drops of water,
Which distil in rain from His vapour :
Which the skies pour down
And drop upon man abundantly."

Throughout the books of Holy Scripture, the connection between the clouds and the rain is clearly borne in mind. Deborah says in her song "the clouds dropped water." In the Psalms there are many references. In lxxvii. 17, "The clouds poured out water;" in cxlvii. 8, "Who covereth the heaven with clouds, Who prepareth rain for the earth." Proverbs xvi. 15, "His favour is as a cloud of the latter rain." The Preacher says that "clouds return after the rain"; and Isaiah, "I will also command the clouds that they rain no rain upon it"; and Jude, "Clouds they are without water, carried about of winds."

The clouds, too, were not conceived as being heavy. Nahum says that "the clouds are the dust of His feet," and Isaiah speaks of "a cloud of dew in the heat of harvest." The Preacher clearly understood that "the waters above" were not pent in by solid barriers; that they were carried by the clouds; for "if the clouds be full of rain, they empty themselves upon the earth." And Job says of Jehovah, "He bindeth up the waters in His thick clouds, and the cloud is not rent under them;" and, later, Jehovah Himself asks:—

> "Canst thou lift up thy voice to the clouds,
> That abundance of waters may cover thee?
> . . . .
> Who can number the clouds by wisdom,
> Or who can pour out the bottles of heaven?"

The Hebrews, therefore, were quite aware that the waters of the sea were drawn up into the atmosphere by evaporation, and were carried by it in the form of clouds. No doubt their knowledge in this respect, as in others, was the growth of time. But there is no need to suppose that, even in the earlier stages of their development, the Hebrews thought of the "waters that be above the heavens" as contained in a literal cistern overhead. Still less is there reason to adopt Prof. Schiaparelli's strange deduction: "Considering the spherical and convex shape of the firmament, the upper waters could not remain above without a second wall to hold them in at the sides and the top. So a second vault above the vault of the firmament closes in, together with the firmament, a space where are the storehouses of rain, hail, and snow."[1] There seems to be nowhere in Scripture the slightest hint or suggestion of any such second vault; certainly not in the beautiful passage to which Prof. Schiaparelli is here referring.

[1] *Astronomy in the Old Testament*, p. 32.

"Where is the way to the dwelling of light,
  And as for darkness, where is the place thereof;
  That thou shouldest take it to the bound thereof,
  And that thou shouldst discern the paths to the house thereof.
          .   .   .   .
  Hast thou entered the treasuries of the snow,
  Or hast thou seen the treasuries of the hail,
  Which I have reserved against the time of trouble,
  Against the day of battle and war?
  By what way is the light parted,
  Or the east wind scattered upon the earth?
  Who hath cleft a channel for the water-flood,
  Or a way for the lightning of the thunder;
          .   .   .   .
          Hath the rain a father?
  Or who hath begotten the drops of dew?
  Out of whose womb came the ice?
  And the hoary frost of heaven, who hath gendered it?"

The Song of David, Psalm xviii., clearly shows that its writer held no fantasy of a solidly built cistern of waters in the sky, but thought of the "dark waters" in the heavens, as identical with the "thick clouds." The passage is worth quoting at some length, not merely as supplying a magnificent word picture of a storm, but as showing the free and courageous spirit of the Hebrew poet, a spirit more emancipated than can be found in any other nation of antiquity. It was not only the gentler aspect of nature that attracted him; even for its most terrible, he had a sympathy, rising, under the influence of his strong faith in God, into positive exultation in it.

"In my distress I called upon the Lord,
  And cried unto my God:
  He heard my voice out of His temple,
  And my cry before Him came into His ears.

> Then the earth shook and trembled,
> The foundations also of the mountains moved
> And were shaken, because He was wroth.
> There went up a smoke out of His nostrils,
> And fire out of His mouth devoured:
> Coals were kindled by it.
> He bowed the heavens also, and came down;
> And thick darkness was under His feet.
> And He rode upon a cherub, and did fly:
> Yea, He flew swiftly upon the wings of the wind.
> He made darkness His hiding place,
> His pavilion round about Him;
> Darkness of waters, thick clouds of the skies.
> At the brightness before Him His thick clouds passed,
> Hailstones and coals of fire.
> The Lord also thundered in the heavens,
> And the Most High uttered His voice;
> Hailstones and coals of fire.
> And He sent out His arrows, and scattered them;
> Yea lightnings manifold, and discomfited them.
> Then the channels of waters appeared,
> And the foundations of the world were laid bare,
> At Thy rebuke, O Lord,
> At the blast of the breath of Thy nostrils.
> He sent from on high, He took me;
> He drew me out of many waters.
> He delivered me from my strong enemy,
> And from them that hated me, for they were too mighty for me."

Two other passages point to the circulation of water vapour upward from the earth before its descent as rain; one in the prophecy of Jeremiah, the other, almost identical with it, in Psalm cxxxv. 7: " When He uttereth His voice, there is a tumult of waters in the heavens, and He causeth the vapours to ascend from the ends of the earth; He maketh lightnings for the rain, and bringeth forth the wind out of His treasuries." Here we get a hint of a close observing of nature among the Hebrews. For by the foreshortening that clouds undergo in the distance, they inevitably appear to form chiefly on the horizon, "at the ends of the earth," whence they move upwards towards the zenith.

A further reference to clouds reveals not observation only but acute reflection, though it leaves the mystery without solution. "Dost thou know the balancings of the clouds, the wondrous works of Him Which is perfect in knowledge?" There is a deep mystery here, which science is far from having completely solved, how it is that the clouds float, each in its own place, at its own level; each perfectly "balanced" in the thin air.

"That mist which lies in the morning so softly in the valley, level and white, through which the tops of the trees rise as if through an inundation—why is *it* so heavy? and why does it lie so low, being yet so thin and frail that it will melt away utterly into splendour of morning, when the sun has shone on it but a few moments more? Those colossal pyramids, huge and firm, with outlines as of rocks, and strength to bear the beating of the high sun full on their fiery flanks—why are *they* so light—their bases high over our heads, high over the heads of Alps? why will these melt away, not as the sun rises, but as he descends, and leave the stars of twilight clear, while the valley vapour gains again upon the earth like a shroud?"[1]

The fact of the "balancing" has been brought home to us during the past hundred years very vividly by the progress of aërial navigation. Balloons are objects too familiar even to our children to cause them any surprise,

[1] Ruskin, *Modern Painters*, part vii. chap. i.

CIRRUS FROM SOUTH KENSINGTON, 1906, MAY 29.

CUMULI FROM TUNBRIDGE WELLS, 1906, MAY 20.
(Photographs of clouds, taken by Dr. W. J. S. Lockyer.)

"Dost thou know the balancing of the clouds?"

and every one knows how instantly a balloon, when in the air, rises up higher if a few pounds of ballast are thrown out, or sinks if a little of the gas is allowed to escape. We know of no balancing more delicate than this, of a body floating in the air.

"The spreadings of the clouds" mentioned by Elihu are of the same nature as their "balancings," but the expression is less remarkable. The "spreading" is a thing manifest to all, but it required the mind both of a poet and a man of science to appreciate that such spreading involved a delicate poising of each cloud in its place.

The heavy rain which fell at the time of the Deluge is indeed spoken of as if it were water let out of a reservoir by its floodgates;—"the windows of heaven were opened;" but it seems to show some dulness on the part of an objector to argue that this expression involves the idea of a literal stone-built reservoir with its sluices. Those who have actually seen tropical rain in full violence will find the Scriptural phrase not merely appropriate but almost inevitable. The rain does indeed fall like hitherto pent-up waters rushing forth at the opening of a sluice, and it seems unreasonable to try to place too literal an interpretation upon so suitable a simile.

There is the less reason to insist upon this very matter-of-fact rendering of the "windows of heaven," that in two out of the three connections in which it occurs, the expression is certainly used metaphorically. On the occasion of the famine in the city of Samaria, Elisha prophesied that—

"To-morrow about this time shall a measure of fine flour be sold for a shekel, and two measures of barley for a shekel, in the gate of Samaria. Then a lord on whose hand the king leaned answered the man of God, and said, Behold, if the Lord would make windows in heaven, might this thing be?"

So again Malachi exhorted the Jews after the Return from Babylon:—

"Bring ye all the tithes into the storehouse, that there may be meat in Mine house, and prove Me now herewith, saith the Lord of hosts, if I will not open you the windows of heaven, and pour you out a blessing, that there shall not be room enough to receive it."

In neither case can the "windows of heaven" have been meant by the speaker to convey the idea of the sluice-gates of an actual, solidly-built reservoir in the sky.

One other cloud fact—their dissipation as the sun rises high in the heavens—is noticed in one of the most tender and pathetic passages in all the prophetic Scriptures. The Lord, by the mouth of Hosea, is mourning over the instability of His people. "O Ephraim, what shall I do unto thee? O Judah, what shall I do unto thee? For your goodness is as a morning cloud, and as the early dew it goeth away."

The winds of heaven were considered as four in number, corresponding to our own four "cardinal points." Thus the great horn of Daniel's he-goat was broken and succeeded by four notable horns toward the four winds of heaven; as the empire of Alexander the Great was divided amongst his four generals. In Ezekiel's vision of the dry bones the prophet prays, "Come from the four

THE FIRMAMENT

winds, O breath, and breathe upon these slain;" and Jeremiah foretells that "the four winds from the four quarters of heaven" shall be brought upon Elam, and scatter its outcasts into every nation.

The circulation of the winds is clearly set forth by the Preacher in the Book of Ecclesiastes.

> "The wind goeth toward the south, and turneth about unto the north; it whirleth about continually, and the wind returneth again according to his circuits."

Of the four quarters the Hebrews reckoned the east as first. It was to the east that they supposed themselves always looking. The chief word for east, therefore, *kedem*, means "that which is before," "the front"; and the word next in use is, naturally, *mizrach*, the rising of the sun. The west is, as naturally, *mebō hasshemesh*, the going down of the sun; but as the Mediterranean Sea lay to the westward of Palestine "the sea" (*yam*) is frequently put instead of that point of the compass. With the east in front, the south becomes the right, and the north the left. The south also was *negeb*, the desert, since the desert shut in Palestine to the south, as the sea to the west. In opposition to *tsaphon*, the dark or hidden north, the south is *darom*, the bright and sunny region.

The phrase "four corners of the earth" does not imply that the Hebrews thought of the earth as square. Several expressions on the contrary show that they thought of it as circular. The Lord "sitteth upon the circle of the earth," and in another passage the same form is applied to the ocean. "He set a compass (*margin* circle) upon the face of the depth." This circle is no doubt the circle of the visible horizon, within which earth and sea are spread out apparently as a plain; above it "the vault of heaven" (Job xxii. 14; R.V. *margin*) is arched. There does not appear to be allusion, anywhere in Scripture, to the spherical form of the earth.

The Hebrew knowledge of the extent of the terrestrial plain was of course very limited, but it would seem that, like many other nations of antiquity, they supposed that the ocean occupied the outer part of the circle surrounding the land which was in the centre. This may be inferred from Job's statement—

"He hath described a boundary upon the face of the waters,
  Unto the confines of light and darkness."

The boundary of the world is represented as being "described," or more properly "circumscribed," drawn as a circle, upon the ocean. This ocean is considered as essentially one, exactly as by actual exploration we now know it to be;—"Let the waters under the heaven be gathered together unto one place;"—all the oceans and seas communicate.

Beneath the earth there are the waters. The Lord hath founded the world " upon the seas, and established it upon the floods," and (Psalm cxxxvi. 6) "stretched out the earth above the waters." This for the most part means simply that the water surface lies lower than the land surface. But there are waters,—other than those of the ocean,—which are, in a strict sense, beneath the earth; the subterranean waters, which though in the very substance of the earth, and existing there in an altogether different way from the great masses of water we see upon the surface, form a water system, which may legitimately be termed a kind of ocean underground. From these subterranean waters our springs issue forth, and it is these waters we tap in our wells. Of the cedar in Lebanon Ezekiel spoke: "The waters made him great, the deep set him up on high with her rivers running round about his plants, and sent out her little rivers (*margin*, conduits) unto all the trees of the field." The " deep," *tchōm*, applies therefore, not merely to the restless waters of the ocean, but to these unseen waters as well; and means, not merely "surging waters," but depths of any kind. When in the great Deluge the floodgates of heaven were opened, these "fountains of the great deep were broken up" as well. And later both fountains and windows were "stopped." So the Lord asks Job, "Hast thou entered into the springs of the sea? or hast thou walked in the search of the depth?" and in Proverbs it is said of the Lord, "By His knowledge the depths are broken up, and the clouds drop down the dew."

The tides upon the sea-coast of Palestine are very slight, but some have seen a reference to them in Jer. v. 22, where the Lord says, I "have placed the sand for the bound of the sea by a perpetual decree, that it cannot pass it: and though the waves thereof toss themselves, yet can they not prevail; though they roar, yet can they not pass over it." More probably the idea to be conveyed is merely that of the restraint of the sea to its proper basin, as in the passage where the Lord asks Job, "Who shut up the sea with doors, when it brake forth, as if it had issued out of the womb?" And the writer of Proverbs sums all up:—

"When He prepared the heavens, I [Wisdom] was there: when He set a compass upon the face of the depth: when He established the clouds above: when He strengthened the fountains of the deep: when He gave to the sea His decree, that the waters should not pass His commandment: when He appointed the foundations of the earth."

CHAPTER V

THE ORDINANCES OF THE HEAVENS

As has been already pointed out, the astronomical references in Scripture are not numerous, and probably give but an inadequate idea of the actual degree of progress attained by the Hebrews in astronomical science. Yet it is clear, even from the record which we have, that there was one great astronomical fact which they had observed, and that it had made a deep impression upon them.

That fact was the sublime Order of the heavenly movements. First amongst these was the order of the daily progress of the sun; rising in the east and moving slowly, majestically, and resistlessly upward to the meridian,—the "midst" or "bisection" of heaven, of Josh. x. 13,—and then passing downwards as smoothly and unfalteringly to his setting in the west.

This motion of the sun inspires the simile employed by the Psalmist in the astronomical psalm, the nineteenth. He sings—

"The heavens declare the glory of God.
. . . .
In them hath He set a tabernacle for the sun,
Which is as a bridegroom coming out of his chamber,
And rejoiceth as a strong man to run his course.

> His going forth is from the end of the heaven,
> And his circuit unto the ends of it:
> And there is nothing hid from the heat thereof."

The night revealed another Order, in its way more majestic still. As the twilight faded away the bright and silent watchers of the heavens mustered each in his place. And each, like the sun during the day, was moving, slowly, majestically, resistlessly, "without haste, without rest." Each had its appointed place, its appointed path. Some moved in small circles in the north; some rose in the east, and swept in long curves over towards their setting in the west, some scarcely lifted themselves above the southern horizon. But each one kept its own place. None jostled another, or hurried in advance, or lagged behind. It is no wonder that as the multitude of the stars was observed, and the unbroken order of their going, that the simile suggested itself of an army on the march—" the host of heaven." And the sight of the unbroken order of these bright celestial orbs suggested a comparison with the unseen army of exalted beings, the angels; the army or host of heaven in another sense, marshalled, like the stars, in perfect obedience to the Divine will. So in the vision of Micaiah, the son of Imlah, the "host of heaven" are the thousands of attendant spirits waiting around the throne of God to fulfil His bidding.

"I saw the Lord sitting on His throne, and all the host of heaven standing by him on His right hand and on His left."

But more frequently it is the starry, not the angelic, army to which reference is made.

So Jeremiah prophesies—

> "As the host of heaven cannot be numbered,
> Neither the sand of the sea measured:
> So will I multiply the seed of David My servant,
> And the Levites that minister unto Me."

The prophets of Israel recognized clearly, that the starry host of heaven and the angelic host were distinct; that the first, in their brightness, order, and obedience formed fitting comparison for the second; but that both were created beings; neither were divinities.

The heathen nations around recognized also the hosts both of the stars and of spiritual beings, but the first they took as the manifestations of the second, whom they counted as divinities. There was often a great confusion between the two, and the observance or worship of the first could not be kept distinguished from the recognition or worship of the other; the very ideogram for a god was an 8-rayed star.

The Hebrews were warned again and again lest, confusing in their minds these two great hosts of stars and angels, they should deem the one the divine manifestation of the other, the divinity, not accounting them both fellow-servants, the handiwork of God.

Thus, in the wilderness, the Lord commands them through Moses—

> "Take ye therefore good heed unto yourselves, . . . lest thou lift up thine eyes unto heaven, and when thou seest the sun, and the moon, and the stars, even all the host of heaven, shouldest be driven to worship them, and serve them, which the Lord thy God hath divided [distributed] unto all nations under the whole heaven."

But the one celestial army continually suggests the other, and the two are placed in the closest parallelism when reference is made to the time when the foundations of the earth were fastened, and the corner-stone thereof was laid,

> " When the morning stars sang together,
> And all the sons of God shouted for joy."

So when Deborah sings of the deliverance which the Lord gave to Israel at the battle of the Kishon, she puts the stars for the angelic legions that she feels assured were engaged in warring in their support.

> "They fought from heaven;
> The stars in their courses fought against Sisera."

The "courses" of the stars are the paths which they appear to follow as they move round the pole of the heavens as the night proceeds, whilst the stars themselves stand for the heavenly helpers who, unseen, had mingled in the battle and discomforted the squadrons of Sisera's war-chariots. It almost reads as if to Deborah had been vouchsafed such a vision as Elisha prayed might be given to his servant:—

"Therefore sent the King of Syria thither horses, and chariots, and a great host: and they came by night, and compassed the city about.

"And when the servant of the man of God was risen early, and gone forth, behold, an host compassed the city both with horses and chariots. And his servant said unto him, Alas, my master! how shall we do?

"And he answered, Fear not: for they that be with us are more than they that be with them.

"And Elisha prayed, and said, Lord, I pray Thee, open

THE ORDINANCES OF THE HEAVENS

his eyes, that he may see. And the Lord opened the eyes of the young man; and he saw: and, behold, the mountain was full of horses and chariots of fire round about Elisha."

The solemn procession of the starry host through the long night—the rising in the east, the southing, and the setting in the west—is not the only ordered movement of the stars of heaven that may be recognized. As night by night brightens to its dawn, if we watch the eastern horizon and note what stars are the last to rise above it before the growing daylight overpowers the feeble stellar rays, then we see that some bright star, invisible on the preceding mornings, shines out for a few moments low down in the glimmer of the dawn. As morning succeeds morning it rises earlier, until at last it mounts when it is yet dark, and some other star takes its place as the herald of the rising sun. We recognize to-day this "heliacal rising" of the stars. Though we do not make use of it in our system of time-measuring, it played an important part in the calendar-making of the ancients. Such heralds of the rising sun were called "morning stars" by the Hebrews, and they used them "for seasons" and "for years." One star or constellation of stars would herald by its "heliacal rising" the beginning of spring, another the coming of winter; the time to plough, the time to sow, the time of the rains, would all be indicated by the successive "morning stars" as they appeared. And after an interval of three hundred and sixty-five or three hundred and sixty-six days the same star would again show itself as a morning star for a second time, marking out the year, whilst the other morning stars would follow, each in its due season. So we read in Job, that God led "forth the Mazzaroth in their season."

This wonderful procession of the midnight sky is not known and admired by those who live in walled cities and ceiled houses, as it is by those who live in the open, in the wilderness. It is not therefore to be wondered at, that we find praise of these "works of the Lord . . . sought out of all them that have pleasure therein," mostly amongst the shepherds, the herdsmen, the wanderers in the open—in the words and prophecies of Job, of Jacob, Moses, David and Amos.

The thought that each new day, beginning with a new outburst of light, was, in its degree, a kind of new creation, an emblem of the original act by which the world was brought into being, renders appropriate and beautiful the ascription of the term "morning stars" to those "sons of God," the angels. As the stars in the eastern sky are poetically thought of as "singing together" to herald the creation of each new day, so in the verses already quoted from the Book of Job, the angels of God are represented as shouting for joy when the foundations of the earth were laid.

The "morning star" again stands as the type and earnest of that new creation which God has promised to His servants. The epistle to Thyatira concludes with the promise—" He that overcometh, and keepeth my works unto the end, . . . I will give him the morning star."

The brightest of these heralds of the sun is the planet Venus, and such a "morning star" for power, glory, and

THE ORDINANCES OF THE HEAVENS 61

magnificence, the king of Babylon had once been; like one of the angels of God. But as addressed in Isaiah's prophecy, he has been brought down to Sheol:—

"How art thou fallen from heaven, O Lucifer, son of the morning! . . . For thou hast said in thine heart, I will ascend into heaven, I will exalt my throne above the stars of God . . . I will ascend above the heights of the clouds; I will be like the most High."

But the "morning star" is taken as a higher type, even of our Lord Himself, and of His future coming in glory. St. Peter bids the disciples, to whom he writes, take heed unto the word of prophecy as unto a lamp shining in a dark place "until the day dawn, and the Day star arise in your hearts." In almost the last words of the Bible, the Lord uses the same image Himself:—

"I, Jesus, have sent Mine angel to testify unto you these things in the Churches. I am the root and the offspring of David, the bright and morning star."

In the sublime and ordered movements of the various heavenly bodies, the Hebrews recognized the ordinances of God. The point of view always taken in Scripture is the theo-centric one; the relation sought to be brought out is not the relation of thing to thing—which is the objective of physical science—but the relation of creature to Creator. We have no means of knowing whether they made attempt to find any mechanical explanation of the movements; such inquiry would lie entirely outside the scope of the books of Holy Scripture, and other ancient Hebrew literature has not been transmitted to us.

The lesson which the Psalmists and the Prophets desired to teach was not the daily rotation of the earth upon its axis, nor its yearly revolution round the sun, but that—

"If those ordinances depart from before Me, saith the Lord, then the seed of Israel also shall cease from being a nation before Me for ever."

In the Bible all intermediate steps are omitted, and the result is linked immediately to the first Cause. God Himself is the theme, and trust in Him the lesson.

"Lift up your eyes on high, and see Who hath created these, That bringeth out their host by number: He calleth them all by name; by the greatness of His might, and for that He is strong in power, not one is lacking.

"Why sayest thou, O Jacob, and speakest, O Israel, My way is hid from the Lord, and my judgment is passed away from my God. Hast thou not known? hast thou not heard? the everlasting God, the Lord, the Creator of the ends of the earth, fainteth not, neither is weary; there is no searching of His understanding. He giveth power to the faint; and to him that hath no might He increaseth strength. Even the youths shall faint and be weary, and the young men shall utterly fall: but they that wait upon the Lord shall renew their strength; they shall mount up with wings as eagles; they shall run, and not be weary; they shall walk, and not faint."

CHAPTER VI

THE SUN

"AND God said, Let there be lights in the firmament of the heaven to divide the day from the night; and let them be for signs, and for seasons, and for days, and years: and let them be for lights in the firmament of the heaven, to give light upon the earth: and it was so. And God made two great lights; the greater light to rule the day, and the lesser light to rule the night: He made the stars also. And God set them in the firmament of the heaven to give light upon the earth, and to rule over the day and over the night, and to divide the light from the darkness: and God saw that it was good. And the evening and the morning were the fourth day."

A double purpose for the two great heavenly bodies is indicated here,—first, the obvious one of giving light; next, that of time-measurement. These, from the human and practical point of view, are the two main services which the sun and moon render to us, and naturally sufficed for the object that the writer had before him. There is no evidence that he had any idea that the moon simply shone by reflecting the light of the sun; still less that the sun was a light for worlds other than our own; but if he had known these facts we can hardly suppose that he would have mentioned them; there would have been no purpose to be served by so doing.

But it is remarkable that no reference is made either to the incalculable benefits conferred by the action of the sun in ripening the fruits of the earth, or to the services of the moon as a time-measurer, in dividing off the months. Both these actions are clearly indicated later on in the Scriptures, where Moses, in the blessing which he pronounced upon the tribe of Joseph, prayed that his land might be blessed "for the precious things of the fruits of the sun," so that we may take their omission here, together with the omission of all mention of the planets, and the slight parenthetical reference to the stars, as indicating that this chapter was composed at an exceedingly early date.

The chief purpose of the sun is to give light; it "rules" or regulates the day and "divides the light from the darkness." As such it is the appropriate emblem of God Himself, Who "is Light, and in Him is no darkness at all." These images are frequently repeated in the Scriptures, and it is only possible to give a few instances. David sings, "The Lord is my light and my salvation." "The Lord shall be unto thee an everlasting light," is the promise made to Zion. St. John expressly uses the term of the Son of God, our Lord: "That was the true Light which lighteth every man that cometh into the world." Whilst the more concrete emblem is used as often. In the eighty-fourth psalm, the psalm of pilgrimage, we read, "The Lord God is a sun and shield;" Malachi predicts that "the Sun of Righteousness shall arise with healing in His wings," and St. James, with the same thought of the sun in his mind, speaks of God as "the Father of lights."

But in none of these or the other parallel passages is there the remotest approach to any deification of the sun, or even of that most ethereal of influences, light itself. Both are creatures, both are made by God; they are things and things only, and are not even the shrines of a deity. They may be used as emblems of God in some of His attributes; they do not even furnish any indication of His special presence, for He is equally present where sun and light are not. "The darkness hideth not from Thee; but the night shineth as the day: the darkness and the light are both alike to Thee."

The worship of the sun and of other heavenly bodies is one of the sins most unsparingly denounced in Scripture. It was one of the first warnings of the Book of Deuteronomy that Israel as a people were to take heed "lest thou lift up thine eyes unto heaven, and when thou seest the sun, and the moon, and the stars, even all the host of heaven, shouldest be driven to worship them and serve them," and the utter overthrow of the nation was foretold should they break this law. And as for the nation, so for the individual, any "man or woman that hath wrought wickedness in the sight of the Lord thy God, in transgressing His covenant, and hath gone and served other gods, and worshipped them, either the sun, or moon, or any of the host of heaven" was when convicted of working "such abomination" unsparingly to be put to death.

Yet with all this, sun-worship prevailed in Israel again

and again. Two of the reforming kings of Judah, Asa and Josiah, found it necessary to take away "the sun-images;" indeed, the latter king found that the horses and chariots which his predecessors, Manasseh and Amon, had dedicated to sun-worship, were kept at the very entrance to the temple. In spite of his reformation, however, the evil spread until the final corruption of Jerusalem was shown in vision to Ezekiel; "Seventy men of the ancients"—that is, the complete Sanhedrim—offered incense to creeping things and abominable beasts; the women wept for Tammuz, probably the sun-god in his decline to winter death; and, deepest apostasy of all, five and twenty men, the high-priest, and the chief priests of the twenty-four courses, "with their backs toward the temple of the Lord, and their faces toward the east; and they worshipped the sun toward the east." The entire nation, as represented in its chief members in State, Society, and Church, was apostate, and its ruin followed. Five years more and the temple was burned and Jerusalem destroyed, and in captivity and exile the nation learned to abhor the idolatry that had brought about its overthrow.

Four words are translated "sun" in our Authorized Version. Of these one, used Job xxxi. 26, should really be "light," as in the margin,—"If I beheld the light when it shined,"—though the sun is obviously meant. The second word is one used in poetry chiefly in conjunction with a poetical word for the moon, and refers to the sun's warmth, as the other does to the whiteness of the moon. Thus the Bride in the Song of Solomon is described as "fair as the moon, clear as the sun." The third word has given rise to some ambiguity. In the eighth chapter of Judges, in the Authorized Version, it is stated that "Gideon, the son of Joash, returned from the battle before the sun was up," but in the Revised Version that he "returned from the battle from the ascent of Heres." There was a mount Heres, a mount of the sun, in the portion of the Danites held by the Amorites, but that cannot have been the Heres of Gideon. Still the probability is that a mount sacred to the sun is meant here as well as in the reference to the Danites; though *heres* as meaning the sun itself occurs in the story of Samson's riddle, for the men of the city gave him the answer to it which they had extorted from his wife, "before the sun (*heres*) went down." *Shemesh*, the *Samas* of the Babylonians, is the usual word for the sun; and we find it in Beth-shemesh, the "house of the sun," a Levitical city within the tribe of Judah, the scene of the return of the ark after its captivity amongst the Philistines. There was another Beth-shemesh in Naphtali on the borders of Issachar, and Jeremiah prophesies that Nebuchadnezzar "shall break also the images of Beth-shemesh, that is in the land of Egypt," probably the obelisks of the sun in On, or Heliopolis. It was from this city that Joseph, when vizier of Egypt, took his wife, the daughter of the high priest there. The images of the sun, and of Baal as the sun-god, seem to have been obelisks or pillars of stone, and hence had to be "broken down"; whilst the Asherah, the "groves" of the Authorized Version, the images of Ashtoreth as the moon-goddess, were wooden pillars, to be "cut" or "hewn down."

Another "city of the sun" in the land of Egypt is also mentioned by Isaiah, in his prophecy of the conversion and restoration of the Egyptians. "Five cities in the land of Egypt shall speak the language of Canaan, and swear to the Lord of hosts; one shall be called The city of destruction;" lit. of *Ḥeres*, or of the sun. It was upon the strength of this text that Onias, the son of Onias the high priest, appealed to Ptolemy Philometer to be allowed to build a temple to Jehovah in the prefecture of Heliopolis (the city of the sun), and obtained his permission to do so, B.C. 149.[1]

The epithet applied to the sun in Cant. vi. already quoted, "Clear as the sun," may be taken as equivalent to "spotless." That is its ordinary appearance to the naked eye, though from time to time—far more frequently than most persons have any idea—there are spots upon the sun sufficiently large to be seen without any optical assistance. Thus in the twenty years from 1882 to 1901 inclusive, such a phenomenon occurred on the average once in each week. No reference to the existence of sun-spots occurs in Scripture. Nor is this surprising, for it would not have fallen within the purpose of Scripture to record such a fact. But it is surprising that whilst the Chinese detected their occasional appearance, there is no distinct account of such an observation given either on Babylonian tablets or by classical or mediæval writers.

The achievement of the Chinese in this direction is very notable, for the difficulty of looking directly at the

[1] Josephus, *Antiquities*, XIII. iii. 1.

sun, under ordinary circumstances, is so great, and the very largest sunspots are so small as compared with the entire disc, that it argues great perseverance in watching such appearances on the part of the Chinese, for them to have assured themselves that they were not due to very small distant clouds in our own atmosphere.

It has often been the subject of comment that light is mentioned in Gen. i. as having been created on the first day, but the sun not until the fourth. The order is entirely appropriate from an astronomical point of view, for we know that our sun is not the only source of light, since it is but one out of millions of stars, many of which greatly exceed it in splendour. Further, most astronomers consider that our solar system existed as a luminous nebula long ages before the sun was formed as a central condensation.

But the true explanation of the creation of light being put first is probably this—that there might be no imagining that, though gross solid bodies, like earth and sea, sun and moon might require a Creator, yet something so ethereal and all-pervading as light was self-existent, and by its own nature, eternal. This was a truth that needed to be stated first. God is light, but light is not God.

The other references to the sun in Scripture do not call for much comment. Its apparent unchangeableness qualifies it for use as an expression for eternal duration, as in the seventy-second, the Royal, Psalm, "They shall fear Thee as long as the sun and moon endure;" and again, "His name shall endure for ever: His name shall be continued as long as the sun." And again, in the eighty-ninth Psalm, it is said of David: "His seed shall endure for ever, and his throne as the sun before Me."

The daily course of the sun from beyond the eastern horizon to beyond the western gives the widest expression for the compass of the whole earth. "The mighty God, even the Lord, hath spoken, and called the earth, from the rising of the sun unto the going down thereof." "From the rising of the sun, unto the going down of the same, the Lord's name is to be praised." The sun's rays penetrate everywhere. "His going forth is from the end of the heaven, and his circuit unto the ends of it: and there is nothing hid from the heat thereof." Whilst in the Book of Ecclesiastes, the melancholy words of the Preacher revert over and over again to that which is done "under the sun." "What profit hath a man of all his labour which he taketh under the sun?"

It should be noted that this same Book of Ecclesiastes shows a much clearer idea of the sun's daily apparent motion than was held by many of the writers of antiquity. There is, of course, nowhere in Scripture any mention of the rotation of the earth on its axis as the mechanical explanation of the sun's daily apparent motion; any more than we should refer to it ourselves to-day except when writing from a purely technical point of view. As said already, the Hebrews had probably not discovered this explanation, and would certainly have not gone out of their way to mention it in any of their Scriptures if they had.

One passage of great beauty has sometimes been quoted as if it contained a reference to the earth's rotation, but when carefully examined it is seen to be dealing simply with the apparent motion of the sun in the course of the year and of the day.

> "Hast thou commanded the morning since thy days;
> And caused the dayspring to know his place;
> That it might take hold of the ends of the earth,
> That the wicked might be shaken out of it?
> It is turned as clay to the seal;
> And they stand as a garment."

The earth appears to be spoken of as being "turned" to the sun, the dayspring; and this, we know, takes place, morning by morning, in consequence of the diurnal rotation. But the last two lines are better rendered in the Revised Version—

> "It is changed as clay under the seal;
> And *all things* stand forth as a garment."

The ancient seals were cylinders, rolled over the clay, which, formless before, took upon it the desired relief as the seal passed over it. So a garment, laid aside and folded up during the night, is shapeless, but once again takes form when the wearer puts it on. And the earth, formless in the darkness, gains shape and colour and relief with the impress upon it of the morning light.

It is quite clear that the Hebrews did not suppose that it was a new sun that came up from the east each morning, as did Xenophanes and the Epicureans amongst the Greeks. It was the same sun throughout. Nor is there any idea of his hiding himself behind a mysterious mountain during the night. "The sun," the Preacher tells us, "ariseth and the sun goeth down, and hasteth to his place where he arose." The Hebrew was quite aware that the earth was unsupported in space, for he knew that the Lord "stretcheth out the north over the empty place, and hangeth the earth upon nothing." There was therefore nothing to hinder the sun passing freely under the earth from west to east, and thus making his path, not a mere march onward ending in his dissolution at sunset, but a complete "circuit," as noted by the writer of the nineteenth Psalm.

The fierceness of the sun's heat in Palestine rendered sun-stroke a serious danger. The little son of the Shunammite was probably so smitten as he watched his father at work with the reapers. So the promise is given to God's people more than once: "The sun shall not smite thee by day." "They shall not hunger nor thirst; neither shall the heat nor sun smite them." The martyrs who pass through the great tribulation "shall hunger no more, neither thirst any more; neither shall the sun light on them, nor any heat."

There are fewer references in Scripture to the vivifying effects of sunlight upon vegetation than we might have expected. The explanation is possibly to be found in the terrible perversion men had made of the benefits which came to them by means of this action of sunlight, by using them as an excuse for plunging into all kinds of nature-worship. Yet there are one or two allusions not without interest. As already mentioned, "the precious fruits brought forth by the sun" were promised to the tribe of Joseph, whilst the great modern discovery that nearly every form of terrestrial energy is derived ultimately from the energy of the sun's rays gives a most striking appropriateness to the imagery made use of by St. James.

"Every good gift and every perfect gift is from above, and cometh down from the Father of Lights, with Whom is no variableness, neither shadow of turning."

God, that is to say, is the true Sun, the true Origin of all Lights, the true bestower of every good and perfect gift. The word rendered "variableness," is a technical word, used by ourselves in modern English as "parallax," and employed in the Septuagint Version to denote the revolutions of the heavenly bodies, described in the thirty-eighth chapter of the book of Job, as "the ordinances of the heavens." With the natural sun, therefore, there is "variableness," that is to say, real or apparent change of place; there is none with God. Neither is there with Him any darkness of eclipse; any "shadow" caused as in the case of the material sun, by the "turning" of earth and moon in their orbits. The knowledge of "the alternations of the turning of the sun," described in the Book of Wisdom as a feature of the learning of Solomon, was a knowledge of the laws of this "variableness" and "turning"; especially of the "turning" of its rising and setting points at the two solstices; and St. James may well have had that passage in his mind when he wrote. For Science deals with the knowledge of things that change, as they change, and of their changes, but Faith with the knowledge of Him that abideth for ever, and it is to this higher knowledge that St. James wished to point his readers.

Science deals with the knowledge of things that change, as they change and of their changes. The physical facts that we have learned in the last years about that changeful body the sun are briefly these :—

Its core or inner nucleus is not accessible to observation, its nature and constitution being a mere matter of inference. The "photosphere" is a shell of incandescent cloud surrounding the nucleus, but the depth, or thickness of this shell is quite unknown. The outer surface—which we see —of the photosphere is certainly pretty sharply defined, though very irregular, rising at points into whiter aggregations, called "faculæ," and perhaps depressed at other places in the dark "spots." Immediately above the photosphere lies the "reversing layer" in which are found the substances which give rise to the gaps in the sun's spectrum —the Fraunhofer lines. Above the "reversing layer" lies the scarlet "chromosphere" with "prominences" of various forms and dimensions rising high above the solar surface; and over, and embracing all, is the "corona," with its mysterious petal-like forms and rod-like rays.

The great body of the sun is gaseous, though it is impossible for us to conceive of the condition of the gaseous core, subjected, as it is, at once to temperature and pressure both enormously great. Probably it is a gas so viscous that it would resist motion as pitch or putty does. Nor do we know much of the nature of either the sun-spots or the solar corona. Both seem to be produced by causes which lie within the sun; both undergo changes that are periodical and connected with each other. They exercise some influence upon the earth's magnetism, but whether this influence extends to terrestrial weather, to rainfall and storms, is still a matter of controversy.

THE SUN

The sun itself is distant from the earth in the mean, about 92,885,000 miles, but this distance varies between January and June by 3,100,000 miles. The diameter of the sun is 866,400 miles, but perhaps this is variable to the extent of some hundreds of miles. It would contain 1,305,000 times the bulk of the earth, but its mean density is but one-quarter that of the earth. The force of gravity at its surface is $27\frac{1}{2}$ times that at the surface of the earth, and it rotates on its axis in about 25 days. But the sun's surface does not appear to rotate as a whole, so this time of rotating varies by as much as two days if we consider a region on the sun's equator or at a distance from it of 45°. The intensity of sunlight at the surface of the sun is about 190,000 times that of a candle-flame, and the effective temperature of the solar surface is eight or ten thousand degrees centigrade.

Such are some of the facts about the sun that are received, or, as it would be technically expressed, "adopted" to-day. Doubtless a very few years will find them altered and rendered more accurate as observations accumulate. In a few hundred years, knowledge of the constitution of the sun may have so increased that these data and suggestions may seem so erroneous as to be absurd. It is little more than a century since one of the greatest of astronomers, Sir William Herschel, contended that the central globe of the sun might be a habitable world, sheltered from the blazing photosphere by a layer of cool non-luminous clouds. Such an hypothesis was not incompatible with what was then known of the constitution of the heavenly bodies, though it is incompatible with what we know

now. It was simply a matter on which more evidence was to be accumulated, and the holding of such a view does not, and did not, detract from the scientific status of Sir William Herschel.

The hypotheses of science require continual restatement in the light of new evidence, and, as to the weight and interpretation to be given to such evidence, there is continual conflict—if it may so be called—between the old and the new science, between the science that is established and the science that is being established. It is by this conflict that knowledge is rendered sure.

Such a conflict took place rather more than 300 years ago at the opening of the Modern Era of astronomy. It was a conflict between two schools of science—between the disciples of Aristotle and Claudius Ptolemy on the one hand and the disciples of Copernicus on the other. It has often been represented as a conflict between religion and science, whereas that which happened was that the representatives of the older school of science made use of the powers of the Church to persecute the newer school as represented by Galileo. That persecution was no doubt a flagrant abuse of authority, but it should be impossible at the present day for any one to claim a theological standing for either theory, whether Copernican or Ptolemaic.

So long as evidence sufficient to demonstrate the Copernican hypothesis was not forthcoming, it was possible for a man to hold the Ptolemaic, without detracting from his scientific position, just as it is thought no discredit to Sir William Herschel that he held his curious idea of a cool sun under the conditions of knowledge of a hundred

THE SUN

years ago. Even at the present day, we habitually use the Ptolemaic phraseology. Not only do we speak of "sunrise" and "sunset," but astronomers in strictly technical papers use the expression, "acceleration of the sun's motion" when "acceleration of the earth's motion" is meant.

The question as to whether the earth goes round the sun or the sun goes round the earth has been decided by the accumulation of evidence. It was a question for evidence to decide. It was an open question so long as the evidence available was not sufficient to decide it. It was perfectly possible at one time for a scientific or a religious man to hold either view. Neither view interfered with his fundamental standing or with his mental attitude towards either sun or earth. In this respect—important as the question is in itself—it might be said to be a mere detail, almost a matter of indifference.

But it is not a mere detail, a matter of indifference to either scientist or religious man, as to what the sun and earth *are*—whether he can treat them as things that can be weighed, measured, compared, analyzed, as, a few pages back, we have shown has been done, or whether, as one of the chief astrologers of to-day puts it, he—

"Believes that the sun is the body of the Logos of this solar system, 'in Him we live and move and have our being.' The planets are his angels, being modifications in the consciousness of the Logos,"

and that the sun

"Stands as Power, having Love and Will united."

The difference between these two points of view is fundamental, and one of root principle. The foundation, the common foundation on which both the believer and the scientist build, is threatened by this false science and false religion. The calling, the very existence of both is assailed, and they must stand or fall together. The believer in one God cannot acknowledge a Sun-god, a Solar Logos, these planetary angels; the astronomer cannot admit the intrusion of planetary influences that obey no known laws, and the supposed effects of which are in no way proportional to the supposed causes. The Law of Causality does not run within the borders of astrology.

It is the old antithesis restated of the Hebrew and the heathen. The believer in one God and the scientist alike derive their heritage from the Hebrew, whilst the modern astrologer claims that the astrology of to-day is once more a revelation of the Chaldean and Assyrian religions. But polytheism—whether in its gross form of many gods, of planetary angels, or in the more subtle form of pantheism,—is the very negation of sane religion; and astrology is the negation of sane astronomy.

"For the invisible things of Him from the creation of the world are clearly seen, being understood by the things that are made, even His eternal power and Godhead; so that they are without excuse: because that, when they knew God, they glorified Him not as God, neither were thankful; but became vain in their imaginations, and their foolish heart was darkened. Professing themselves to be wise, they became fools, and changed the glory of the uncorruptible God into an image made like to corruptible man, and to birds, and fourfooted beasts, and creeping things."

CHAPTER VII

THE MOON

> "The balmy moon of blessed Israel
> Floods all the deep-blue gloom with beams divine:
> All night the splintered crags that wall the dell
> With spires of silver shine."

So, in Tennyson's words, sang Jephthah's daughter, as she recalled the days of her mourning before she accomplished her self-sacrifice.

It is hard for modern dwellers in towns to realize the immense importance of the moon to the people of old. "The night cometh when no man can work" fitly describes their condition when she was absent. In sub-tropical countries like Palestine, twilight is short, and, the sun once set, deep darkness soon covers everything. Such artificial lights as men then had would now be deemed very inefficient. There was little opportunity, when once darkness had fallen, for either work or enjoyment.

But, when the moon was up, how very different was the case. Then men might say—

> "This night, methinks, is but the daylight sick;
> It looks a little paler: 'tis a day,
> Such as the day is when the sun is hid."

In the long moonlit nights, travelling was easy and safe; the labours of the field and house could still be carried on; the friendly feast need not be interrupted. But of all men, the shepherd would most rejoice at this season; all his toils, all his dangers were immeasurably lightened during the nights near the full. As in the beautiful rendering which Tennyson has given us of one of the finest passages in the *Iliad*—

> "In heaven the stars about the moon
> Look beautiful, when all the winds are laid,
> And every height comes out, and jutting peak
> And valley, and the immeasurable heavens
> Break open to their highest, and all the stars
> Shine, and the Shepherd gladdens in his heart."

A large proportion of the people of Israel, long after their settlement in Palestine, maintained the habits of their forefathers, and led the shepherd's life. To them, therefore, the full of the moon must have been of special importance; yet there is no single reference in Scripture to this phase as such; nor indeed to any change of the moon's apparent figure. In two cases in our Revised Version we do indeed find the expression "at the full moon," but if we compare these passages with the Authorized Version, we find them there rendered "in the time appointed," or "at the day appointed." This latter appears to be the literal meaning, though there can be no question, as is seen by a comparison with the Syriac, that the period of the full moon is referred to. No doubt it was because travelling was so much more safe and easy than in the moonless nights, that the two great spring and

THE MOON 81

autumn festivals of the Jews were held at the full moon. Indeed, the latter feast, when the Israelites "camped out" for a week "in booths," was held at the time of the "harvest moon." The phenomenon of the "harvest moon" may be briefly explained as follows. At the autumnal equinox, when the sun is crossing from the north side of the equator to the south, the full moon is crossing from the south side of the equator to the north. It is thus higher in the sky, when it souths, on each succeeding night, and is therefore up for a greater length of time. This counterbalances to a considerable extent its movement eastward amongst the stars, so that, for several nights in succession, it rises almost at sundown. These nights of the Feast of Tabernacles, when all Israel was rejoicing over the ingathered fruits, each family in its tent or arbour of green boughs, were therefore the fullest of moonlight in the year.[1]

Modern civilization has almost shut us off from the heavens, at least in our great towns and cities. These offer many conveniences, but they remove us from not a few of the beauties which nature has to offer. And so it comes that, taking the population as a whole, there is perhaps less practically known of astronomy in England to-day than there was under the Plantagenets. A very few are astronomers, professional and amateur, and know immeasurably more than our forefathers did of the science. Then there is a large, more or less cultured, public that know something of the science at secondhand through

[1] How the little children must have revelled in that yearly holiday!

books. But the great majority know nothing of the heavenly bodies except of the sun; they need to "look in the almanack" to "find out moonshine." But to simpler peoples the difference between the "light half" of the month, from the first quarter to the last quarter through the full of the moon, and the "dark half," from the last quarter to the first quarter, through new, is very great. Indian astronomers so divide the month to this day.

In one passage of Holy Scripture, the description which Isaiah gives of the "City of the Lord, the Zion of the Holy One of Israel," there is a reference to the dark part of the month.

"Thy sun shall no more go down; neither shall thy moon (literally "month") withdraw itself: for the Lord shall be thine everlasting light, and the days of thy mourning shall be ended."

The parallelism expressed in the verse lies between the darkness of night whilst the sun is below the horizon, and the special darkness of those nights when the moon, being near conjunction with the sun, is absent from the sky during the greater part or whole of the night hours, and has but a small portion of her disc illuminated. Just as half the day is dark because the sun has withdrawn itself, so half the nights of the month are dark because the moon has withdrawn itself.

. The Hebrew month was a natural one, determined by actual observation of the new moon. They used three words in their references to the moon, the first of which, *chodesh*, derived from a root meaning "to be new," indicates the fact that the new moon, as actually observed, governed their calendar. The word therefore signifies the new moon—the day of the new moon: and thus a month; that is, a lunar month beginning at the new moon. This is the Hebrew word used in the Deluge story in the seventh chapter of Genesis; and in all references to feasts depending on a day in the month. As when the Lord spake to Moses, saying, "Also in the day of your gladness, and in your solemn days, and in the beginnings of your months, ye shall blow with your trumpets over your burnt offerings, and over the sacrifices of your peace offerings." And again in the Psalm of Asaph to the chief musician upon Gittith: "Blow up the trumpet in the new moon, in the time appointed, on our solemn feast day." This is the word also that Isaiah uses in describing the bravery of the daughters of Zion, "the tinkling ornaments about their feet, and their cauls, and their round tires like the moon, the chains, and the bracelets." "The round tires" were not discs, like the full moon, but were round like the crescent.

Generally speaking, *chodesh* is employed where either reference is made to the shape or newness of the crescent moon, or where "month" is used in any precise way. This is the word for "month" employed throughout by the prophet Ezekiel, who is so precise in the dating of his prophecies.

When the moon is mentioned as the lesser light of heaven, without particular reference to its form, or when a month is mentioned as a somewhat indefinite period of time, then the Hebrew word *yarēach*, is used. Here the word has the root meaning of "paleness"; it is the "silver moon."

Yarēach is the word always used where the moon is classed among the heavenly bodies; as when Joseph dreamed of the sun, the moon, and the eleven constellations; or in Jer. viii. 2, where the Lord says that they shall bring out the bones of the kings, princes, priests, prophets, and inhabitants of Jerusalem, "and they shall spread them before the sun, and the moon, and all the host of heaven, whom they have loved, and whom they have served, and after whom they have walked, and whom they have sought, and whom they have worshipped."

The same word is used for the moon in its character of "making ordinances." Thus we have it several times in the Psalms: "He (the Lord) appointed the moon for seasons." "His seed shall endure for ever, and his throne as the sun before Me. It shall be established for ever as the moon, and as a faithful witness in heaven." And again: "The moon and stars rule by night;" whilst Jeremiah says, "Thus saith the Lord, Which giveth the sun for a light by day, and the ordinances of the moon and of the stars for a light by night."

In all passages where reference seems to be made to the darkening or withdrawing of the moon's light (Eccl. xii. 2; Isa. xiii. 10; Ezek. xxxii. 7; Joel ii. 10, 31, and iii. 15; and Hab. iii. 11) the word *yarēach* is employed. A slight variant of the same word indicates the month when viewed as a period of time not quite defined, and not in the strict sense of a lunar month. This is the term used in Exod. ii. 2, for the three months that the mother

THE MOON

of Moses hid him when she saw that he was a goodly child; by Moses, in his prophecy for Joseph, of "Blessed of the Lord be his land . . . for the precious fruits brought forth by the sun, and for the precious things put forth by the months." Such a "full month of days" did Shallum the son of Jabesh reign in Samaria in the nine and thirtieth year of Uzziah, king of Judah. Such also were the twelve months of warning given to Nebuchadnezzar, king of Babylon, before his madness fell upon him. The same word is once used for a true lunar month, viz. in Ezra vi. 15, when the building of the "house was finished on the third day of the month Adar, which was in the sixth year of the reign of Darius the king." In all other references to the months derived from the Babylonians, such as the "month Chisleu" in Neh. i. 1, the term *chodesh* is used, since these, like the Hebrew months, were defined by the observation of the new moon; but for the Tyrian months, Zif, Bul, Ethanim, we find the term *yerach* in three out of the four instances.

In three instances a third word is used poetically to express the moon. This is *lebanah*, which has the meaning of whiteness. In Song of Sol. vi. 10, it is asked—

"Who is she that looketh forth as the morning, fair as the moon, clear as the sun, and terrible as an army with banners?"

Isaiah also says—

"Then the moon shall be confounded, and the sun ashamed, when the Lord of Hosts shall reign in Mount Zion, and in Jerusalem, and before His ancients gloriously."

And yet again—

"Moreover the light of the moon shall be as the light of the sun, and the light of the sun shall be sevenfold, as the light of seven days, in the day that the Lord bindeth up the breach of His people, and healeth the stroke of their wound."

It may not be without significance that each of these three passages, wherein the moon is denominated by its name of whiteness or purity, looks forward prophetically to the same great event, pictured yet more clearly in the Revelation—

"And I heard as it were the voice of a great multitude, and as the voice of many waters, and as the voice of mighty thunderings, saying, Alleluia: for the Lord God omnipotent reigneth.
"Let us be glad and rejoice, and give honour to Him: for the marriage of the Lamb is come, and His wife hath made herself ready.
"And to her was granted that she should be arrayed in fine linen, clean and white: for the fine linen is the righteousness of saints."

Chodesh and *yarēach* are masculine words; *lebanah* is feminine. But nowhere throughout the Old Testament is the moon personified, and in only one instance is it used figuratively to represent a person. This is in the case of Jacob's reading of Joseph's dream, already referred to, where he said—

"Behold I have dreamed a dream more; and, behold, the sun and the moon and the eleven stars made obeisance to me."

And his father quickly rebuked him, saying—

"What is this dream that thou hast dreamed? Shall I and thy mother and thy brethren indeed come to bow down ourselves to thee to the earth?"

Here Jacob understands that the moon (*yarēach*) stands for a woman, his wife. But in Mesopotamia, whence his grandfather Abraham had come out, Sin, the moon-god, was held to be a male god, high indeed among the deities at that time, and superior even to Samas, the sun-god. Terah, the father of Abraham, was held by Jewish tradition to have been an especial worshipper of the moon-god, whose great temple was in Haran, where he dwelt.

Wherever the land of Uz may have been, at whatever period Job may have lived, there and then it was an iniquity to worship the moon or the moon-god. In his final defence to his friends, when the "three men ceased to answer Job, because he was righteous in his own eyes," Job, justifying his life, said—

"If I beheld the sun when it shined,
Or the moon walking in brightness;
And my heart hath been secretly enticed,
And my mouth hath kissed my hand:
This also were an iniquity to be punished by the judges:
For I should have lied to God that is above."

The Hebrews, too, were forbidden to worship the sun, the moon, or the stars, the host of heaven, and disobeyed the commandment both early and late in their history. When Moses spake unto all Israel on this side Jordan in the wilderness in the plain over against the Red Sea, he said to them—

"The Lord spake unto you out of the midst of the fire: ye heard the voice of the words, but saw no similitude; only ye heard a voice. . . . Take ye therefore good heed unto yourselves; for ye saw no manner of *similitude* on the day that the Lord spake unto you in Horeb out of the midst of the fire:

"Lest ye corrupt yourselves, and make you a graven image, the similitude of any figure, the likeness of male or female . . . . And lest thou lift up thine eyes unto heaven, and when thou seest the sun, and the moon, and the stars, even all the host of heaven, shouldst be driven to worship them, and serve them, which the Lord thy God hath divided unto all nations under the whole heaven."

We know what the "similitude" of the sun and the moon were like among the surrounding nations. We see their "hieroglyphs" on numberless seals and images from the ruins of Nineveh or Babylon. That of the sun was first a rayed star or disc, later a figure, rayed and winged. That of the moon was a crescent, one lying on its back, like a bowl or cup, the actual attitude of the new moon at the beginning of the new year. Just such moon similitudes did the soldiers of Gideon take from off the camels of Zebah and Zalmunna; just such were the "round tires like the moon" that Isaiah condemns among the bravery of the daughters of Zion.

The similitude or token of Ashtoreth, the paramount goddess of the Zidonians, was the *ashera*, the "grove" of the Authorized Version, probably in most cases merely a wooden pillar. This goddess, "the abomination of the Zidonians," was a moon-goddess, concerning whom Eusebius preserves a statement by the Phœnician historian, Sanchoniathon, that her images had the head of an ox. In the wars in the days of Abraham we find Chedorlaomer, and the kings that were with him, smiting the Rephaim in Ashteroth Karnaim, that is, in the Ashtoreths "of the horns." It is impossible to decide at this date whether the horns which gave the distinctive title to this shrine of Ashtoreth owed their origin to the horns of the animal merged in the goddess, or to the horns of the crescent moon, with which she was to some extent identified. Possibly there was always a confusion between the two in the minds of her worshippers. The cult of Ashtoreth was spread not only among the Hebrews, but throughout the whole plain of Mesopotamia. In the times of the Judges, and in the days of Samuel, we find continually the statement that the people "served Baalim and Ashtaroth"—the plurals of Baal and Ashtoreth—these representing the son and moon, and reigning as king and queen in heaven. When the Philistines fought with Saul at Mount Gilboa, and he was slain, they stripped off his armour and put it "in the house of Ashtaroth." Yet later we find that Solomon loved strange women of the Zidonians, who turned his heart after Ashtoreth, the goddess of the Zidonians, and he built a high place for her on the right hand of the Mount of Olives, which remained for some three and a half centuries, until Josiah, the king, defiled it. Nevertheless, the worship of Ashtoreth continued, and the prophet Jeremiah describes her cult:—

"The children gather wood, and the fathers kindle the fire, and the women knead their dough, to make cakes to the queen of heaven."

This was done in the cities of Judah and streets of Jerusalem, but the Jews carried the cult with them even when they fled into Egypt, and whilst there they answered Jeremiah—

"We will certainly do whatsoever thing goeth forth out of our own mouth, to burn incense unto the queen of heaven, and to pour out drink offerings unto her, as we have done, we, and our fathers, our kings, and our princes, in the cities of Judah, and in the streets of Jerusalem: for then had we plenty of victuals, and were well, and saw no evil. But since we left off to burn incense to the queen of heaven, and to pour out drink offerings unto her, we have wanted all things, and have been consumed by the sword and by the famine."

Ashtoreth, according to Pinches [1] is evidently a lengthening of the name of the Assyrio-Babylonian goddess Ištar, and the Babylonian legend of the Descent of Ištar may well have been a myth founded on the varying phases of the moon. But it must be remembered that, though Ashtoreth or Ištar might be the queen of heaven, the moon was not necessarily the only aspect in which her worshippers recognized her. In others, the planet Venus may have been chosen as her representative; in others the constellation Taurus, at one time the leader of the Zodiac; in others, yet again, the actual form of a material bull or cow.

The Hebrews recognized the great superiority in brightness of the sun over the moon, as testified in their names of the "greater" and "lesser" lights, and

[1] T. G. Pinches, *The Old Testament in the Light of the Historical Records of Assyria and Babylonia*, p. 278.

in such passages as that already quoted from Isaiah (xxx. 26). The word here used for moon is the poetic one, *lebanah*. Of course no argument can be founded on the parallelism employed so as to lead to the conclusion that the Hebrews considered that the solar light exceeded the lunar by only seven times, instead of the 600,000 times indicated by modern photometric measurement.

In only one instance in Scripture — that already quoted of the moon withdrawing itself—is there even an allusion to the changing phases of the moon, other than that implied in the frequent references to the new moons. The appointment of certain feasts to be held on the fifteenth day of the month is a confirmation of the supposition that their months were truly lunar, for then the moon is fully lighted, and rides the sky the whole night long from sunset to sunrise. It is clear, therefore, that the Hebrews, not only noticed the phases of the moon, but made regular use of them. Yet, if we adopted the argument from silence, we should suppose that they had never observed its changes of shape, for there is no direct allusion to them in Scripture. We cannot, therefore, argue from silence as to whether or no they had divined the cause of those changes, namely that the moon shines by reflecting the light of the sun.

Nor are there any references to the markings on the moon. It is quite obvious to the naked eye that there are grey stains upon her silver surface, that these grey stains are always there, most of them forming a chain which curves through the upper hemisphere. Of the bright parts of the moon, some shine out with greater lustre than others, particularly one spot in the lower left-hand quadrant, not far from the edge of the full disc. The edges of the moon gleam more brightly as a rule than the central parts. All this was apparent to the Hebrews of old, as it is to our unassisted sight to-day.

The moon's influence in raising the tides is naturally not mentioned. The Hebrews were not a seafaring race, nor are the tides on the coast of Palestine pronounced enough to draw much attention. One influence is ascribed to the moon; an influence still obscure, or even disputed. For the promise that—

> "The sun shall not smite thee by day,
> Nor the moon by night,"

quite obvious in its application to the sun, with the moon seems to refer to its supposed influence on certain diseases and in causing "moon-blindness."

The chief function of the moon, as indicated in Scripture, is to regulate the calendar, and mark out the times for the days of solemnity. In the words of the 104th Psalm :—

> "He (God) appointed the moon for seasons:
> The sun knoweth his going down.
> Thou makest darkness, and it is night;
> Wherein all the beasts of the forest do creep forth.
> The young lions roar after their prey,
> And seek their meat from God.
> The sun ariseth, they get them away,
> And lay them down in their dens.
> Man goeth forth unto his work
> And to his labour until the evening.
> O Lord, how manifold are Thy works!
> In wisdom hast Thou made them all :
> The earth is full of Thy riches."

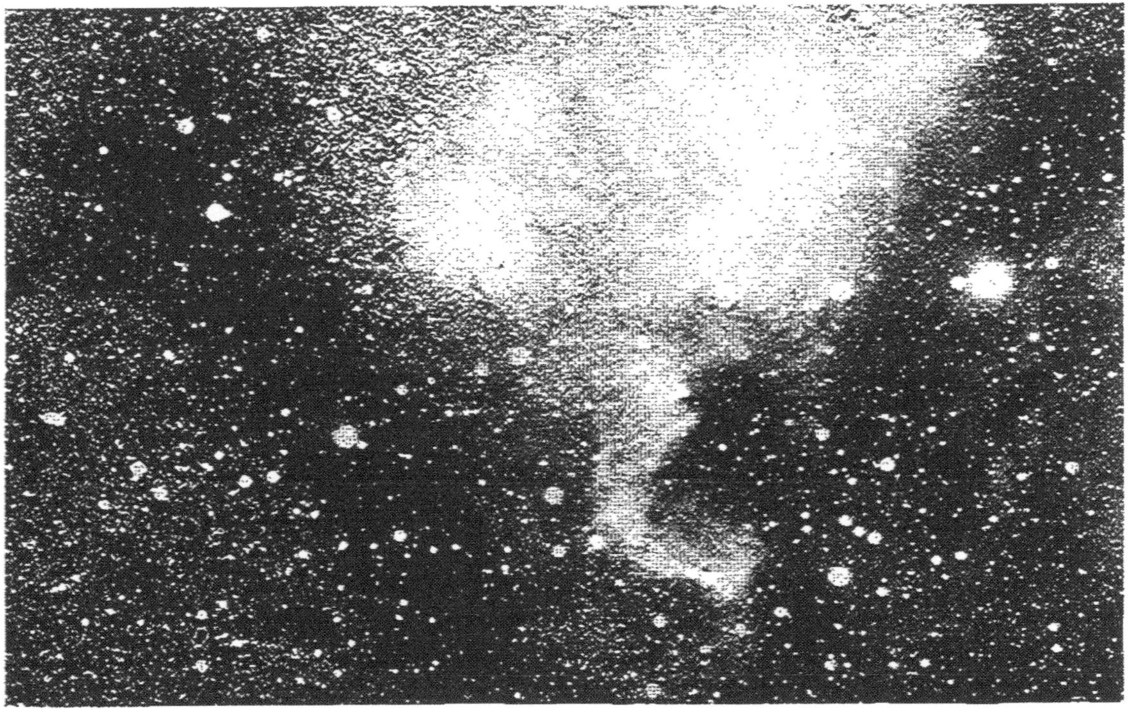

A CORNER OF THE MILKY WAY.
The "America Nebula"; photographed by Dr. Max Wolf, at Heidelberg.

CHAPTER VIII

THE STARS

THE stars and the heaven, whose host they are, were used by the Hebrew writers to express the superlatives of number, of height, and of expanse. To an observer, watching the heavens at any particular time and place, not more than some two thousand stars are separately visible to the unassisted sight. But it was evident to the Hebrew, as it is to any one to-day, that the stars separately visible do not by any means make up their whole number. On clear nights the whole vault of heaven seems covered with a tapestry or curtain the pattern of which is formed of patches of various intensities of light, and sprinkled upon this patterned curtain are the brighter stars that may be separately seen. The most striking feature in the pattern is the Milky Way, and it may be easily discerned that its texture is made up of innumerable minute points of light, a granulation, of which some of the grains are set more closely together, forming the more brilliant patches, and some more loosely, giving the darker shades. The mind easily conceives that the minute points of light whose aggregations make up the varying pattern of the Milky Way, though too small to be individually seen, are also stars, differing perhaps from the stars of the Pleiades or the Bears only in their greater distance or smaller size. It was of all these that the Lord said to Abram—

"Look now toward heaven, and tell the stars, if thou be able to number them: and He said unto him, So shall thy seed be."

The first catalogue of the stars of which we have record was that of Hipparchus in 129 B.C. It contained 1,025 stars, and Ptolemy brought this catalogue up to date in the Almagest of 137 A.D. Tycho Brahé in 1602 made a catalogue of 777 stars, and Kepler republished this in 1627, and increased the number to 1,005. These were before the invention of the telescope, and consequently contained only naked-eye stars. Since astronomers have been able to sound the heavens more deeply, catalogues have increased in size and number. Flamsteed, the first Astronomer Royal, made one of 3,310 stars; from the observations of Bradley, the third, a yet more famous catalogue has been compiled. In our own day more than three hundred thousand stars have been catalogued in the Bonn Durchmusterung; and the great International Photographic Chart of the Heavens will probably show not less than fifty millions of stars, and in this it has limited itself to stars exceeding the fourteenth magnitude in brightness, thus leaving out of its pages many millions of stars that are visible through our more powerful telescopes.

So when Abraham, Moses, Job or Jeremiah speaks of the host of heaven that cannot be numbered, it does not mean simply that these men had but small powers of numeration. To us,—who can count beyond that which we can conceive,—as to the Psalmist, it is a sign of infinite power, wisdom and knowledge that "He telleth the number of the stars; He calleth them all by their names."

Isaiah describes the Lord as "He that sitteth upon the circle of the earth, . . . that stretcheth out the heavens as a curtain, and spreadeth them out as a tent to dwell in." And many others of the prophets use the same simile of a curtain which we have seen to be so appropriate to the appearance of the starry sky. Nowhere, however, have we any indication whether or not they considered the stars were all set *on* this curtain, that is to say were all at the same distance from us. We now know that they are not equidistant from us, but this we largely base on the fact that the stars are of very different orders of brightness, and we judge that, on an average, the fainter a star appears, the further is it distant from us. To the Hebrews, as to us, it was evident that the stars differ in magnitude, and the writer of the Epistle to the Corinthians expressed this when he wrote—

"There is one glory of the sun, and another glory of the moon, and another glory of the stars: for one star differeth from another star in glory."

The ancient Greek astronomers divided the stars according to their brightness into six classes, or six "magnitudes," to use the modern technical term. The average star of any particular magnitude gives about two and a half times as much light as the average star of the next magnitude. More exactly, the average first magnitude star gives one hundred times the light of the average star of the sixth magnitude.

In a few instances we have been able to measure, in the very roughest degree, the distances of stars; not a hundred stars have their parallaxes known, and these have all been measured in the course of the last century. And so far away are these stars, even the nearest of them, that we do not express their distance from us in millions of miles; we express it in the time that their light takes in travelling from them to us. Now it takes light only one second to traverse 186,300 miles, and yet it requires four and a third years for the light from the nearest star to reach us. This is a star of the first magnitude, Alpha in the constellation of the Centaur. The next nearest star is a faint one of between the seventh and eighth magnitudes, and its light takes seven years to come. From a sixth magnitude star in the constellation of the Swan, the light requires eight years; and from Sirius, the brightest star in the heavens, light requires eight and a half years. These four stars are the nearest to us; from no other star, that we know of, does light take less than ten years to travel; from the majority of those whose distance we have succeeded in measuring, the light takes at least twenty years.

To get some conception of what a "light-year" means, let us remember that light could travel right round the earth at its equator seven times in the space of a single second, and that there are 31,556,925 seconds in a year. Light then could girdle the earth a thousand million times whilst it comes from Alpha Centauri. Or we may put it another way. The distance from Alpha Centauri exceeds the equator of the earth by as much as this exceeds an inch and a half; or by as much as the distance from London to Manchester exceeds the hundredth of an inch.

Of all the rest of the innumerable stars, as far as actual measurement is concerned, for us, as for the Hebrews, they might all actually lie on the texture of a curtain, at practically the same distance from us.

We have measured the distances of but a very few stars; the rest—as every one of them was for the Hebrew—are at a greater distance than any effort of ours can reach, be our telescopes ever so great and powerful, our measuring instruments ever so precise and delicate. For them, as for us, the heaven of stars is "for height," for a height which is beyond measure and therefore the only fitting image for the immensity of God.

So Zophar the Naamathite said—

"Canst thou find out the Almighty unto perfection?
It is as high as heaven; what canst thou do?"

and Eliphaz the Temanite reiterated still more strongly—

"Is not God in the height of heaven?
And behold the height of the stars, how high they are."

God Himself is represented as using the expanse of heaven as a measure of the greatness of his fidelity and mercy. The prophet Jeremiah writes—

"Thus saith the Lord; if Heaven above can be measured, and the foundations of the earth searched out beneath, I will also cast off all the seed of Israel for all that they have done, saith the Lord."

As if he were using the figure of a great cross, whose height was that of the heavens, whose arms stretched from east to west, David testifies of the same mercy and forgiveness :—

"For as the heaven is high above the earth,
So great is His mercy toward them that fear Him.
As far as the east is from the west,
So far hath He removed our transgressions from us."

THE GREAT COMET OF 1843.
"Running like a road through the constellations" (see p. 105);

CHAPTER IX

COMETS

GREAT comets are almost always unexpected visitors. There is only one great comet that we know has been seen more than once, and expect with reasonable certainty to see again. This is Halley's comet, which has been returning to a near approach to the sun at somewhat irregular intervals of seventy-five to seventy-eight years during the last centuries: indeed, it is possible that it was this comet that was coincident with the invasion of England by William the Conqueror.

There are other small comets that are also regular inhabitants of the solar system; but, as with Halley's comet, so with these, two circumstances are to be borne in mind. First, that each successive revolution round the sun involves an increasing degradation of their brightness, since there is a manifest waste of their material at each near approach to the sun; until at length the comet is seen no more, not because it has left the warm precincts of the sun for the outer darkness, but because it has spent its substance. Halley's comet was not as brilliant or as impressive in 1835 as it was in 1759: in 1910 it may have become degraded to an appearance of quite the second order.

Next, we have no knowledge, no evidence, that any of these comets have always been members of the solar family. Some of them, indeed, we know were adopted into it by the influence of one or other of the greater planets: Uranus we know is responsible for the introduction of one, Jupiter of a considerable number. The vast majority of comets, great or small, seem to blunder into the solar system anyhow, anywhere, from any direction: they come within the attractive influence of the sun; obey his laws whilst within that influence; make one close approach to him, passing rapidly across our sky; and then depart in an orbit which will never bring them to his neighbourhood again. Some chance of direction, some compelling influence of a planet that it may have approached, so modified the path of Halley's comet when it first entered the solar system, that it has remained a member ever since, and may so remain until it has ceased to be a comet at all.

It follows, therefore, that, as to the number of great comets that may be seen in any age, we can scarcely even apply the laws of probability. During the last couple of thousand years, since chronicles have been abundant, we know that many great comets have been seen. We may suppose, therefore, that during the preceding age, that in which the Scriptures were written, there were also many great comets seen, but we do not know. And most emphatically we are not able to say, from our knowledge of comets themselves and of their motions, that in the days of this or that writer a comet was flaming in the sky.

If a comet had been observed in those ages we might not recognize the description of it. Thus in the fourth year of the 101st Olympiad, the Greeks were startled by a celestial portent. They did not draw fine distinctions, and posterity might have remained ignorant that the terrifying object was possibly a comet, had not Aristotle, who saw it as a boy at Stagira, left a rather more scientifically worded description of it. It flared up from the sunset sky with a narrow definite tail running "like a road through the constellations." In recent times the great comet of 1843 may be mentioned as having exactly such an appearance.

So we cannot expect to find in the Scriptures definite and precise descriptions that we can recognize as those of comets. At the most we may find some expressions, some descriptions, that to us may seem appropriate to the forms and appearances of these objects, and we may therefore infer that the appearance of a comet may have suggested these descriptions or expressions.

The head of a great comet is brilliant, sometimes star-like. But its tail often takes on the most impressive appearance. Donati's comet, in 1858, assumed the most varied shapes—sometimes its tail was broad, with one bright and curving edge, the other fainter and finer, the whole making up a stupendous semi-circular blade-like object. Later, the tail was shaped like a scimitar, and later again, it assumed a duplex form.

Though the bulk of comets is huge, they contain extraordinarily little substance. Their heads must contain some solid matter, but it is probably in the form of a loose aggregation of stones enveloped in vaporous material. There is some reason to suppose that comets are apt to shed some of these stones as they travel along their paths, for the orbits of the meteors that cause some of our greatest "star showers" are coincident with the paths of comets that have been observed.

But it is not only by shedding its loose stones that a comet diminishes its bulk; it loses also through its tail. As the comet gets close to the sun its head becomes heated, and throws off concentric envelopes, much of which consists of matter in an extremely fine state of division. Now it has been shown that the radiations of the sun have the power of repelling matter, whilst the sun itself attracts by its gravitational force. But there is a difference in the action of the two forces. The light-pressure varies with the surface of the particle upon which it is exercised; the gravitational attraction varies with the mass or volume. If we consider the behaviour of very small particles, it follows that the attraction due to gravitation (depending on the volume of the particle) will diminish more rapidly than the repulsion due to light-pressure (depending on the surface of the particle), as we decrease continually the size of the particle, since its volume diminishes more rapidly than its surface. A limit therefore will be reached below which the repulsion will become greater than the attraction. Thus for particles less than the $\frac{1}{85000}$ part of an inch in diameter the repulsion of the sun is greater than its attraction. Particles in the outer envelope of the comet below this size will be driven away in a continuous stream, and will form that thin, luminous fog which we see as the comet's tail.

We cannot tell whether such objects as these were present to the mind of Joel when he spoke of "blood and fire and pillars of smoke"; possibly these metaphors are better explained by a sand- or thunder-storm, especially when we consider that the Hebrew expression for the "pillars of smoke" indicates a resemblance to a palm-tree, as in the spreading out of the head of a sand- or thunder-cloud in the sky. The suggestion has been made,—following the closing lines of *Paradise Lost* (for Milton is responsible for many of our interpretations of Scripture)

> "High in front advanced,
> The brandished sword of God before them blazed,
> Fierce as a comet,"

—that a comet was indeed the "flaming sword which turned every way, to keep the way of the tree of life." There is less improbability in the suggestion made by several writers that, when the pestilence wasted Jerusalem, and David offered up the sacrifice of intercession in the threshing floor of Ornan the Jebusite, the king may have seen, in the scimitar-like tail of a comet such as Donati's, God's "minister,"—"a flame of fire,"—"the angel of the Lord stand between the earth and the heaven, having a drawn sword in his hand stretched out over Jerusalem."

The late R. A. Proctor describes the wanderings of a comet thus:—

"A comet is seen in the far distant depths of space as a faint and scarcely discernible speck. It draws nearer and nearer with continually increasing velocity, growing continually larger and brighter. Faster and faster it rushes on until it makes its nearest approach to our sun, and then, sweeping round him, it begins its long return voyage into infinite space. As it recedes it grows fainter and fainter, until at length it passes beyond the range of the most powerful telescopes made by man, and is seen no more. It has been seen for the first and last time by the generation of men to whom it has displayed its glories. It has been seen for the first and last time by the race of man itself." [1]

"These are . . . wandering stars, to whom is reserved the blackness of darkness for ever."

[1] R. A. Proctor, *The Expanse of Heaven*, p. 134.

FALL OF AN AEROLITE.
"There fell a great star from heaven, burning as it were a lamp," (*see* p. 116).

CHAPTER X

METEORS

GREAT meteorites—"aerolites" as they are called—are like great comets, chance visitors to our world. Now and then they come, but we cannot foretell their coming. Such an aerolite exploded some fifteen miles above Madrid at about $9^h\ 29^m$, on the morning of February 10, 1896:—

"A vivid glare of blinding light was followed in $1\frac{1}{2}$ minutes by a loud report, the concussion being such as not merely to create a panic, but to break many windows, and in some cases to shake down partitions. The sky was clear, and the sun shining brightly, when a white cloud, bordered with red, was seen rushing from south-west to north-east, leaving behind it a train of fine white dust. A red-tinted cloud was long visible in the east."

Many fragments were picked up, and analyzed, and, like other aerolites, were found to consist of materials already known on the earth. The outer crust showed the signs of fire,—the meteoric stone had been fused and ignited by its very rapid rush through the air—but the interior was entirely unaffected by the heat. The manner in which the elements were combined is somewhat peculiar to aerolites; the nearest terrestrial affinity of the minerals aggregated in them, is to be found in the volcanic products from great depths. Thus aerolites seem to be broken-up fragments from the interior parts of globes like our own. They do not come from our own volcanoes, for the velocities with which they entered our atmosphere prove their cosmical origin. Had our atmosphere not entangled them, many, circuiting the sun in a parabolic or hyperbolic curve, would have escaped for ever from our system. The swift motions, which they had on entering our atmosphere, are considerably greater on the average than those of comets, and probably their true home is not in our solar system, but in interstellar space.

The aerolites that reach the surface are not always exploded into very small fragments, but every now and then quite large masses remain intact. Most of these are stony; some have bits of iron scattered through them; others are almost pure iron, or with a little nickel alloy, or have pockets in them laden with stone. There are hundreds of accounts of the falls of aerolites during the past 2,500 years. The Greeks and Romans considered them as celestial omens, and kept some of them in temples. One at Mecca is revered by the faithful Mohammedans, and Jehangir, the great Mogul, is said to have had a sword forged from an iron aerolite which fell in 1620 in the Panjab. Diana of Ephesus stood on a shapeless block which, tradition says, was a meteoric stone, and reference may perhaps be found to this in the speech of the town-clerk of the city to appease the riot stirred up against St. Paul by Demetrius the silversmith and his companions:—

METEORS

"Ye men of Ephesus, what man is there who knoweth not how that the city of the Ephesians is temple-keeper of the great Diana, and of the image which fell down from Jupiter?"

Aerolites come singly and unexpectedly, falling actually to earth on land or sea. "Shooting stars" come usually in battalions. They travel together in swarms, and the earth may meet the same swarm again and again. They are smaller than aerolites, probably mere particles of dust, and for the most part are entirely consumed in our upper atmosphere, so that they do not actually reach the earth. The swarms travel along paths that resemble cometary orbits; they are very elongated ellipses, inclined at all angles to the plane of the ecliptic. Indeed, several of the orbits are actually those of known comets, and it is generally held that these meteorites or "shooting stars" are the *débris* that a comet sheds on its journey.

We can never see the same "shooting star" twice; its visibility implies its dissolution, for it is only as it is entrapped and burnt up in our atmosphere that we see it, or can see it. Its companions in a great meteoric swarm, are, however, as the sand on the sea-shore, and we recognize them as members of the same swarm by their agreement in direction and date. The swarms move in a closed orbit, and it is where this orbit intersects that of the earth that we get a great "star shower," if both earth and swarm are present together at the intersection. If the swarm is drawn out, so that many meteorites are scattered throughout the whole circuit of its orbit, then we get a "shower" every year. If the meteor swarm is more condensed, so as to form a cluster, then the "shower" only comes when the "gem of the ring," as it is termed, is at the intersection of the orbits, and the earth is there too.

Such a conjunction may present the most impressive spectacle that the heavens can afford. The Leonid meteor shower is, perhaps, the most famous. It has been seen at intervals of about thirty-three years, since early in the tenth century. When Ibrahim ben Ahmed lay dying, in the year 902 A.D., it was recorded that "an infinite number of stars were seen during the night, scattering themselves like rain to the right and left, and that year was known as the year of stars." When the earth encountered the same system in 1202 A.D. the Mohammedan record runs that "on the night of Saturday, on the last day of Muharram, stars shot hither and thither in the heavens, eastward and westward, and flew against one another, like a scattering swarm of locusts, to the right and left." There are not records of all the returns of this meteoric swarm between the thirteenth century and the eighteenth, but when the earth encountered it in 1799, Humboldt reported that "from the beginning of the phenomenon there was not a space in the firmament equal in extent to three diameters of the moon that was not filled every instant with bolides and falling stars;" and Mr. Andrew Ellicott, an agent of the United States, cruising off the coast of Florida, watched this same meteoric display, and made the drawing reproduced on the opposite page. In 1833 a planter in South Carolina wrote of a return of this same system, "Never did rain fall much thicker than the meteors fell towards the earth; east, west, north, south,

METEORIC SHOWER OF 1799, NOVEMBER 12.
Seen off Cape Florida, by Mr. Andrew Ellicott.

it was the same." In 1866 the shower was again heavy and brilliant, but at the end of the nineteenth century, when the swarm should have returned, the display was meagre and ineffective.

The Leonid system of meteorites did not always move in a closed orbit round our sun. Tracing back their records and history, we find that in A.D. 126 the swarm passed close to Uranus, and probably at that time the planet captured them for the sun. But we cannot doubt that some such similar sight as they have afforded us suggested the imagery employed by the Apostle St. John when he wrote, "The stars of heaven fell unto the earth, even as a fig-tree casteth her untimely figs, when she is shaken of a mighty wind. And the heavens departed as a scroll when it is rolled together."

And the prophet Isaiah used a very similar figure—

"All the host of heaven shall be dissolved, and the heavens shall be rolled together as a scroll: and all their host shall fall down, as the leaf falleth off from the vine, and as a falling fig from the fig-tree."

Whilst the simile of a great aerolite is that employed by St. John in his description of the star "Wormwood"—

"The third angel sounded, and there fell a great star from heaven, burning as it were a lamp, and it fell upon the third part of the rivers, and upon the fountains of waters."

St. Jude's simile of the "wandering stars, to whom is reserved the blackness of darkness for ever," may have been drawn from meteors rather than from comets. But, as has been seen, the two classes of objects are closely connected.

The word "meteor" is sometimes used for any unusual light seen in the sky. The Zodiacal Light, the pale conical beam seen after sunset in the west in the spring, and before sunrise in the east in the autumn, and known to the Arabs as the "False Dawn," does not appear to be mentioned in Scripture. Some commentators wrongly consider that the expression, "the eyelids of the morning," occurring twice in the Book of Job, is intended to describe it, but the metaphor does not in the least apply.

The Aurora Borealis, on the other hand, seldom though it is seen on an impressive scale in Palestine, seems clearly indicated in one passage. "Out of the north cometh golden splendour" would well fit the gleaming of the "Northern Lights," seen, as they often are, "as sheaves of golden rays."

CHAPTER XI

ECLIPSES OF THE SUN AND MOON

WE do not know what great comets, or aerolites, or 'star-showers" were seen in Palestine during the centuries in which the books of the Bible were composed. But we do know that eclipses, both of the sun and moon, must have been seen, for these are not the results of chance conjunctions. We know more, that not only partial eclipses of the sun, but total eclipses, fell within the period so covered.

There is no phenomenon of nature which is so truly impressive as a total eclipse of the sun. The beautiful pageants of the evening and the morning are too often witnessed to produce the same effect upon us, whilst the storm and the earthquake and the volcano in eruption, by the confusion and fear for personal safety they produce, render men unfit to watch their developments. But the eclipse awes and subdues by what might almost be called moral means alone: no noise, no danger accompanies it; the body is not tortured, nor the mind confused by the rush of the blast, the crash of the thunder-peal, the rocking of the earthquake, or the fires of the volcano. The only sense appealed to is that of sight; the movements of the orbs of heaven go on without noise or confusion, and with a majestic smoothness in which there is neither hurry nor delay.

This impression is felt by every one, no matter how perfectly acquainted, not only with the cause of the phenomenon, but also with the appearances to be expected, and scientific men have found themselves awestruck and even overwhelmed.

But if such are the feelings called forth by an eclipse now-a-days, in those who are expecting it, who are prepared for it, knowing perfectly what will happen and what brings it about, how can we gauge aright the unspeakable terror such an event must have caused in ages long ago, when it came utterly unforeseen, and it was impossible to understand what was really taking place?

And so, in olden time, an eclipse of the sun came as an omen of terrible disaster, nay as being itself one of the worst of disasters. It came so to all nations but one. But to that nation the word of the prophet had come—

"Learn not the way of the heathen, and be not dismayed at the signs of heaven; for the heathen are dismayed at them."

God did not reveal the physical explanation of the eclipse to the Hebrews: that, in process of time, they could learn by the exercise of their own mental powers. But He set them free from the slavish fear of the heathen; they could look at all these terror-striking signs without fear; they could look with calmness, with confidence, because they looked in faith.

120 THE ASTRONOMY OF THE BIBLE

It is not easy to exaggerate the advantage which this must have given the Hebrews over the neighbouring nations, from a scientific point of view. The word of God gave them intellectual freedom, and so far as they were faithful to it, there was no hindrance to their fully working out the scientific problems which came before them. They neither worshipped the heavenly bodies nor were dismayed at their signs. We have no record as to how far the Hebrews made use of this freedom, for, as already pointed out, the Holy Scriptures were not written to chronicle their scientific achievements. But there can be no doubt that, given the leisure of peace, it is *a priori* more likely that they should have taught astronomy to their neighbours, than have learnt it even from the most advanced.

There must have been numberless eclipses of the moon seen in the ages during which the Canon of Holy Scripture was written. Of eclipses of the sun, total or very nearly total over the regions of Palestine or Mesopotamia, in the times of the Old Testament, we know of four that were actually seen, whose record is preserved in contemporaneous history, and a fifth that was nearly total in Judæa about midday.

The first of the four is recorded on a tablet from Babylon, lately deciphered, in which it states that on "the 26th day of Sivan, day was turned into night, and fire appeared in the midst of heaven." This has been identified with the eclipse of July 31, 1063 B.C., and we do not find any reference to it in Scripture.

The second is that of Aug. 15, 831 B.C. No specific

ECLIPSES OF THE SUN AND MOON

record of this eclipse has been found as yet, but it took place during the lifetime of the prophets Joel and Amos, and may have been seen by them, and their recollection of it may have influenced the wording of their prophecies.

The third eclipse is recorded on a tablet from Nineveh, stating the coincidence of an eclipse in Sivan with a revolt in the city of Assur. This has been identified with the eclipse of June 15, 763 B.C.

The fourth is that known as the eclipse of Larissa on May 18, 603 B.C., which was coincident with the final overthrow of the Assyrian Empire, and the fifth is that of Thales on May 28, 585 B.C.

The earth goes round the sun once in a year, the moon goes round the earth once in a month, and sometimes the three bodies are in one straight line. In this case the intermediate body—earth or moon—deprives the other, wholly or partially of the light from the sun, thus causing an eclipse. If the orbits of the earth and moon were in the same plane, an eclipse would happen every time the moon was new or full; that is to say, at every conjunction and every opposition, or about twenty-five times a year. But the plane of the moon's orbit is inclined to the plane of the earth's orbit at an angle of about 5°, and so an eclipse only occurs when the moon is in conjunction or opposition and is at the same time at or very near one of the nodes—that is, one of the two points where the plane of the earth's orbit intersects the moon's orbit. If the moon is in opposition, or "full," then, under these conditions, an eclipse of the moon takes place, and this is visible at all places where the moon is above the horizon at the time. If, however, the moon is in conjunction, or "new," it is the sun that is eclipsed, and as the shadow cast by the moon is but small, only a portion of the earth's surface will experience the solar eclipse. The nodes of the moon's orbit are not stationary, but have a daily retrograde motion of $3' 10\cdot64''$. It takes the moon therefore 27^d 5^h 5^m 36^s ($27\cdot21222^d$) to perform a journey in its orbit from one node back to that node again; this is called a Draconic period. But it takes the moon 29^d 12^h 44^m $2\cdot87^s$ ($29\cdot53059^d$) to pass from new to new, or from full to full, *i.e.* to complete a lunation. Now 242 Draconic periods very nearly equal 223 lunations, being about 18 years $10\frac{1}{3}$ days, and both are very nearly equal to 19 returns of the sun to the moon's node; so that if the moon is new or full when at a node, in 18 years and 10 or 11 days it will be at that node again, and again new or full, and the sun will be also present in very nearly its former position. If, therefore, an eclipse occurred on the former occasion, it will probably occur on the latter. This recurrence of eclipses after intervals of $18\cdot03$ years is called the Saros, and was known to the Chaldeans. We do not know whether it was known to the Hebrews prior to their captivity in Babylon, but possibly the statement of the wise king, already quoted from the Apocryphal "Wisdom of Solomon," may refer to some such knowledge.

Our calendar to-day is a purely solar one; our months are twelve in number, but of purely arbitrary length, divorced from all connection with the moon; and to us, the Saros cycle does not readily leap to the eye, for eclipses

ECLIPSES OF THE SUN AND MOON

of sun or moon seem to fall haphazard on any day of the month or year.

But with the Hebrews, Assyrians, and Babylonians it was not so. Their calendar was a luni-solar one—their year was on the average a solar year, their months were true lunations; the first day of their new month began on the evening when the first thin crescent of the moon appeared after its conjunction with the sun. This observation is what is meant in the Bible by the "new moon." Astronomers now by "new moon" mean the time when it is actually in conjunction with the sun, and is therefore not visible. Nations whose calendar was of this description were certain to discover the Saros much sooner than those whose months were not true lunations, like the Egyptians, Greeks, and Romans.

There are no direct references to eclipses in Scripture. They might have been used in the historical portions for the purpose of dating events, as was the great earthquake in the days of King Uzziah, but they were not so used. But we find not a few allusions to their characteristic appearances and phenomena in the books of the prophets. God in the beginning set the two great lights in the firmament for signs as well as for seasons; and the prophets throughout use the relations of the sun and moon as types of spiritual relations. The Messiah was the Sun of Righteousness; the chosen people, the Church, was as the moon, which derives her light from Him. The "signs of heaven" were *symbols* of great spiritual events, not *omens* of mundane disasters.

The prophets Joel and Amos are clear and vivid in their

descriptions; probably because the eclipse of 831 B.C. was within their recollection. Joel says first, "The sun and the moon shall be dark;" and again, more plainly,—

"I will show wonders in the heavens and in the earth, blood, and fire, and pillars of smoke. The sun shall be turned into darkness, and the moon into blood, before the great and the terrible day of the Lord come."

This prophecy was quoted by St. Peter on the day of Pentecost. And in the Apocalypse, St. John says that when the sixth seal was opened, "the sun became black as sackcloth of hair, and the moon became as blood."

In these references, the two kinds of eclipses are referred to—the sun becomes black when the moon is "new" and hides it; the moon becomes as blood when it is "full" and the earth's shadow falls upon it; its deep copper colour, like that of dried blood, being due to the fact that the light, falling upon it, has passed through a great depth of the earth's atmosphere. These two eclipses cannot therefore be coincident, but they may occur only a fortnight apart—a total eclipse of the sun may be accompanied by a partial eclipse of the moon, a fortnight earlier or a fortnight later; a total eclipse of the moon may be accompanied by partial eclipses of the sun, both at the preceding and following "new moons."

Writing at about the same period, the prophet Amos says—

"Saith the Lord God, I will cause the sun to go down at noon, and I will darken the earth in the clear day,"

and seems to refer to the fact that the eclipse of 831 B.C. occurred about midday in Judæa.

Later Micah writes—

"The sun shall go down over the prophets, and the day shall be dark over them."

Isaiah says that the "sun shall be darkened in his going forth," and Jeremiah that "her sun is gone down while it was yet day." Whilst Ezekiel says—

"I will cover the sun with a cloud, and the moon shall not give her light. All the bright lights of heaven will I make dark over thee, and set darkness upon thy land, saith the Lord God."

But a total eclipse is not all darkness and terror; it has a beauty and a glory all its own. Scarcely has the dark moon hidden the last thread of sunlight from view, than spurs of rosy light are seen around the black disc that now fills the place so lately occupied by the glorious king of day. And these rosy spurs of light shine on a background of pearly glory, as impressive in its beauty as the swift march of the awful shadow, and the seeming descent of the darkened heavens, were in terror. There it shines, pure, lovely, serene, radiant with a light like molten silver, wreathing the darkened sun with a halo like that round a saintly head in some noble altar-piece; so that while in some cases the dreadful shadow has awed a laughing and frivolous crowd into silence, in others the radiance of that halo has brought spectators to their knees with an involuntary exclamation, "The Glory!" as if God Himself had made known His presence in the moment of the sun's eclipse.

And this, indeed, seems to have been the thought of

both the Babylonians and Egyptians of old. Both nations had a specially sacred symbol to set forth the Divine Presence—the Egyptians, a disc with long outstretched wings; the Babylonians, a ring with wings. The latter symbol on Assyrian monuments is always shown as floating over the head of the king, and is designed to indicate the presence and protection of the Deity.

We may take it for granted that the Egyptians and Chaldeans of old, as modern astronomers to-day, had at

THE ASSYRIAN "RING WITH WINGS."

one time or another presented to them every type of coronal form. But there would, no doubt, be a difficulty in grasping or remembering the irregular details of the corona as seen in most eclipses. Sometimes, however, the corona shows itself in a striking and simple form—when sun-spots are few in number, it spreads itself out in two great equatorial streamers. At the eclipse of Algiers in 1900, already referred to, one observer who watched the eclipse from the sea, said—

"The sky was blue all round the sun, and the effect of

CORONA OF MINIMUM TYPE.

Drawing made by W. H. Wesley, from photographs of the 1900 Eclipse.

the silvery corona projected on it was beyond any one to describe. I can only say it seemed to me what angels' wings will be like." [1]

It seems exceedingly probable that the symbol of the ring with wings owed its origin not to any supposed analogy between the ring and the wings and the divine attributes of eternity and power, but to the revelations of a total eclipse with a corona of minimum type. Moreover the Assyrians, when they insert a figure of their deity within the ring, give him a kilt-like dress, and this kilted or feathered characteristic is often retained where the figure is omitted. This gives the symbol a yet closer likeness to the corona, whose "polar rays" are remarkably like the tail feathers of a bird.

Perhaps the prophet Malachi makes a reference to this characteristic of the eclipsed sun, with its corona like "angels' wings," when he predicts—

"But unto you that fear My name shall the Sun of Righteousness arise with healing in His wings."

But, if this be so, it must be borne in mind that the prophet uses the corona as a simile only. No more than the sun itself, is it the Deity, or the manifestation of the Deity.

In the New Testament we may find perhaps a reference to what causes an eclipse—to the shadow cast by a heavenly body in its revolution—its "turning."

"Every good gift and every perfect boon is from above, coming down from the Father of Lights, with Whom can be no variation, neither shadow that is cast by turning."

[1] *The Total Solar Eclipse of May, 1900*, p. 22.

CHAPTER XII

SATURN AND ASTROLOGY

THE planets, as such, are nowhere mentioned in the Bible. In the one instance in which the word appears in our versions, it is given as a translation of *Mazzaloth*, better rendered in the margin as the "twelve signs or constellations." The evidence is not fully conclusive that allusion is made to any planet, even in its capacity of a god worshipped by the surrounding nations.

Of planets, besides the earth, modern astronomy knows Mercury, Venus, Mars, many planetoids, Jupiter, Saturn, Uranus, and Neptune. And of satellites revolving round planets there are at present known, the moon which owns our earth as primary, two satellites to Mars, seven satellites to Jupiter, ten to Saturn, four to Uranus, and one to Neptune.

The ancients counted the planets as seven, numbering the moon and the sun amongst them. The rest were Mercury, Venus, Mars, Jupiter and Saturn. They recognized no satellites to any planet. We have no evidence that the ancient Semitic nations considered that the moon was more intimately connected with the earth than any of the other six.

But though the planets were sometimes regarded as "seven" in number, the ancients perfectly recognized that the sun and moon stood in a different category altogether from the other five. And though the heathen recognized them as deities, confusion resulted as to the identity of the deity of which each was a manifestation. Samas was the sun-god and Baal was the sun-god, but Samas and Baal, or Bel, were not identical, and both were something more than merely the sun personified. Again, Merodach, or Marduk, is sometimes expressly identified with Bel as sun-god, sometimes with the divinity of the planet Jupiter. Similarly Ashtoreth, or Ištar, is sometimes identified with the goddess of the moon, sometimes with the planet Venus. It would not be safe, therefore, to assume that reference is intended to any particular heavenly body, because a deity is mentioned that has been on occasions identified with that heavenly body. Still less safe would it be to assume astronomical allusions in the description of the qualities or characteristics of that deity. Though Ashtoreth, or Ištar, may have been often identified with the planet Venus, it is ridiculous to argue, as some have done, from the expression "Ashteroth-Karnaim," Ashteroth of "the horns," that the ancients had sight or instruments sufficiently powerful to enable them to observe that Venus, like the moon, had her phases, her "horns." Though Nebo has been identified with the planet Mercury, we must not see any astronomical allusion to its being the nearest planet to the sun in Isaiah's coupling the two together, where he says, "Bel boweth down, Nebo stoopeth."

Isaiah speaks of the King of Babylon—

"How art thou fallen from Heaven, O Lucifer, son of the morning!"

The word here translated Lucifer, "light-bearer," is the word *hēlel* from the root *halal*, and means *spreading brightness*. In the Assyrio-Babylonian, the planet Venus is sometimes termed *Mustēlel*, from the root *ēlil*, and she is the most lustrous of all the "morning stars," of the stars that herald the dawn. But except that her greater brilliancy marks her as especially appropriate to the expression, Sirius or any other in its capacity of morning star would be suitable as an explanation of the term.

St. Peter uses the equivalent Greek expression *Phōsphorus* in his second epistle: "A light that shineth in a dark place, until the day dawn and the day-star" (light-bringer) "arise in your hearts."

Isaiah again says—

"Ye are they that forsake the Lord, that forget My holy mountain, that prepare a table for that Troop, and that furnish the drink offering unto that Number."

"Gad" and "Meni," here literally translated as "Troop" and "Number," are in the Revised Version rendered as "Fortune" and "Destiny." A reference to this god "Meni" has been suggested in the mysterious inscription which the King of Babylon saw written by a hand upon the wall, which Daniel interpreted as "God hath numbered thy kingdom, and brought it to an end." By some commentators Meni is understood to be the planet Venus, and Gad to be Jupiter, for these are associated in

Arabian astrology with Fortune or Fate in the sense of good luck. Or, from the similarity of Meni with the Greek *mēnē*, moon, "that Number" might be identified with the moon, and "that Troop," by analogy, with the sun.

It is more probable, if any astrological deities are intended, that the two little star clusters—the Pleiades and the Hyades—situated on the back and head of the Bull, may have been accounted the manifestations of the divinities which are by their names so intimately associated with the idea of multitude. The number seven has been held a sacred number, and has been traditionally associated with both the little star groups.

In one instance alone does there seem to be any strong evidence that reference is intended to one of the five planets known to the ancients, when worshipped as a god; and even that is not conclusive. The prophet Amos, charging the Israelites with idolatry even in the wilderness, asks—

"Have ye offered unto Me sacrifices and offerings in the wilderness forty years, O house of Israel? But ye have borne the tabernacle of your Moloch and Chiun your images, the star of your god, which ye made to yourselves."

But the Septuagint Version makes the accusation run thus:—

"Ye took up the tabernacle of Moloch and the star of your god Remphan, figures which ye made to worship them."

This was the version which St. Stephen quoted in his defence before the High Priest. It is quite clear that it was star worship to which he was referring, for he prefaces his quotation by saying, " God turned and gave them up to serve the host of heaven, as it is written in the book of the prophets."

The difference between the names " Chiun " and " Remphan " is explained by a probable misreading on the part of the Septuagint translators into the Greek, who seemed to have transcribed the initial of the word as " resh," where it should have been " caph "—" R " instead of " K,"—thus the real word should be transliterated " *Kaivan*," which was the name of the planet Saturn both amongst the ancient Arabs and Syrians, and also amongst the Assyrians, whilst "*Kevan*" is the name of that planet in the sacred books of the Parsees. On the other hand, there seems to be some difficulty in supposing that a deity is intended of which there is no other mention in Scripture, seeing that the reference, both by Amos and St. Stephen, would imply that the particular object of idolatry denounced was one exceedingly familiar to them. Gesenius, therefore, after having previously accepted the view that we have here a reference to the worship of Saturn, finally adopted the rendering of the Latin Vulgate, that the word " Chiun " should be translated " statue " or " image." The passage would then become—

"Ye have borne the booth of your Moloch and the image of your idols, the star of your god which ye made for yourselves."

If we accept the view that the worship of the planet Saturn is indeed referred to, it does not necessarily follow

SATURN AND ASTROLOGY

that the prophet Amos was stating that the Israelites in the wilderness actually observed and worshipped him as such. The prophet may mean no more than that the Israelites, whilst outwardly conforming to the worship of Jehovah, were in their secret desires hankering after Sabæism—the worship of the heavenly host. And it may well be that he chooses Moloch and Saturn as representing the cruellest and most debased forms of heathenism.

The planet Saturn gives its name to the seventh day of our week, "Saturn's day," the sabbath of the week of the Jews, and the coincidence of the two has called forth not a few ingenious theories. Why do the days of our week bear their present names, and what is the explanation of their order?

The late well-known astronomer, R. A. Proctor, gives the explanation as follows:—

"The twenty-four hours of each day were devoted to those planets in the order of their supposed distance from the earth,—Saturn, Jupiter, Mars, the Sun, Venus, Mercury, and the Moon. The outermost planet, Saturn, which also travels in the longest period, was regarded in this arrangement as of chief dignity, as encompassing in his movement all the rest, Jupiter was of higher dignity than Mars, and so forth. Moreover to the outermost planet, partly because of Saturn's gloomy aspect, partly because among half-savage races the powers of evil are always more respected than the powers that work for good, a maleficent influence was attributed. Now, if we assign to the successive hours of a day the planets as above-named, beginning with Saturn on the day assigned to that powerful deity, it will be found that the last hour of that day will be assigned to Mars—'the lesser infortune,' as Saturn was 'the greater infortune,' of the old system of astrology—and the first hour of the next day to the next planet, the Sun; the day following Saturday would thus be Sunday. The last hour of Sunday would fall to Mercury, and the first of the next to the Moon; so Monday, the Moon's day, follows Sunday. The next day would be the day of Mars, who, in the Scandinavian theology, is represented by Tuisco; so Tuisco's day, or Tuesday (Mardi), follows Monday. Then, by following the same system, we come to Mercury's day (Mercredi), Woden's day, or Wednesday; next to Jupiter's day, Jove's day (Jeudi), Thor's day, or Thursday; to Venus's day, Vendredi (Veneris dies), Freya's day, or Friday, and so to Saturday again. That the day devoted to the most evil and most powerful of all the deities of the Sabdans (*sic*) should be set apart—first as one on which it was unlucky to work, and afterwards as one on which it was held to be sinful to work—was but the natural outcome of the superstitious belief that the planets were gods ruling the fates of men and nations."[1]

This theory appears at first sight so simple, so plausible, that many are tempted to say, "It must be true," and it has accordingly gained a wide acceptance. Yet a moment's thought shows that it makes many assumptions, some of which rest without any proof, and others are known to be false.

When were the planets discovered? Not certainly at the dawn of astronomy. The fixed stars must have become familiar, and have been recognized in their various groupings before it could have been known that there were others that were not fixed,—were "planets," *i. e.* wanderers. Thus, amongst the Greeks, no planet is alluded to by Hesiod, and Homer mentions no planet other than Venus,

[1] R. A. Proctor, *The Great Pyramid*, pp. 274-276.

SATURN AND ASTROLOGY 137

and apparently regarded her as two distinct objects, according as she was seen as a morning and as an evening star. Pythagoras is reputed to have been the first of the Greek philosophers to realize the identity of Phosphorus and Hesperus, that is Venus at her two elongations, so that the Greeks did not know this until the sixth century before our era. We are yet without certain knowledge as to when the Babylonians began to notice the different planets, but the order of discovery can hardly have been different from what it seems to have been amongst the Greeks—that is to say, first Venus as two separate objects, then Jupiter and Mars, and, probably much later, Saturn and Mercury. This last, again, would originally be considered a pair of planets, just as Venus had been. Later these planets as morning stars would be identified with their appearances as evening stars. After this obscurity had been cleared up, there was a still further advance to be made before the astrologers could have adopted their strange grouping of the sun and moon as planets equally with the other five. This certainly is no primitive conception; for the sun and moon have such appreciable dimensions and are of such great brightness that they seem to be marked off (as in the first chapter of Genesis) as of an entirely different order from all the other heavenly bodies. The point in common with the other five planets, namely their apparent periodical movements, could only have been brought out by very careful and prolonged observation. The recognition, therefore, of the planets as being "seven," two of the seven being the sun and moon, must have been quite late in the history of the world. The connection of

138 THE ASTRONOMY OF THE BIBLE

the "seven planets" with the seven days of the week was something much later still. It implies, as we have seen, the adoption of a particular order for the planets, and this order further implies that a knowledge had been obtained of their relative distances, and involves a particular theory of the solar system, that which we now know as the Ptolemaic. It is not the order of the Babylonians, for they arranged them, Moon, Sun, Mercury, Venus, Saturn, Jupiter, Mars.

There are further considerations which show that the Babylonians could not have given these planetary names to the days of the week. The order of the names implies that a twenty-four hour day was used, but the Babylonian hours were twice the length of those which we use; hence there were only twelve of them. Further, the Babylonian week was not a true week running on continuously; it was tied to the month, and hence did not lend itself to such a notation.

But the order adopted for the planets is that current amongst the Greek astronomers of Alexandria, who did use a twenty-four hour day. Hence it was certainly later than 300 B.C. But the Greeks and Egyptians alike used a week of ten days, not of seven. How then did the planetary names come to be assigned to the seven-day week?

It was a consequence of the power which the Jews possessed of impressing their religious ideas, and particularly their observance of the sabbath day, upon their conquerors. They did so with the Romans. We find such writers as Cicero, Horace, Juvenal and others remarking

SATURN AND ASTROLOGY 139

upon the sabbath, and, indeed, in the early days of the Empire there was a considerable observance of it. Much more, then, must the Alexandrian Greeks have been aware of the Jewish sabbath,—which involved the Jewish week,— at a time when the Jews of that city were both numerous and powerful, having equal rights with the Greek inhabitants, and when the Ptolemies were sanctioning the erection of a Jewish temple in their dominions, and the translation of the Jewish Scriptures into Greek. It was after the Alexandrian Greeks had thus learned of the Jewish week that they assigned the planets to the seven days of that week, since it suited their astrological purposes better than the Egyptian week of ten days. That allotment could not possibly have brought either week or sabbath into existence. Both had been recognized many centuries earlier. It was foisted upon that which had already a venerable antiquity. As Professor Schiaparelli well remarks, "we are indebted for these names to mathematical astrology, the false science which came to be formed after the time of Alexander the Great from the strange intermarriage between Chaldean and Egyptian superstitions and the mathematical astronomy of the Greeks."[1]

There is a widespread notion that early astronomy, whether amongst the Hebrews or elsewhere, took the form of astrology; that the fortune-telling came first, and the legitimate science grew out of it. Indeed, a claim is not infrequently made that no small honour is due to the early astrologers, since from their efforts,

[1] G. V. Schiaparelli, *Astronomy in the Old Testament*, p. 137.

the most majestic of all the sciences is said to have arisen.

These ideas are the exact contrary of the truth. Mathematical, or perhaps as we might better call it, planetary astrology, as we have it to-day, concerns itself with the apparent movements of the planets in the sense that it uses them as its material; just as a child playing in a library might use the books as building blocks, piling, it may be, a book of sermons on a history, and a novel on a mathematical treatise. Astrology does not contribute, has not contributed a single observation, a single demonstration to astronomy. It owes to astronomy all that it knows of mathematical processes and planetary positions. In astronomical language, the calculation of a horoscope is simply the calculation of the " azimuths " of the different planets, and of certain imaginary points on the ecliptic for a given time. This is an astronomical process, carried out according to certain simple formulæ. The calculation of a horoscope is therefore a straightforward business, but, as astrologers all admit, its interpretation is where the skill is required, and no real rules can be given for that.

Here is the explanation why the sun and moon are classed together with such relatively insignificant bodies as the other five planets, and are not even ranked as their chief. The ancient astrologer, like the modern, cared nothing for the actual luminary in the heavens; all he cared for was its written symbol on his tablets, and there Sun and Saturn could be looked upon as equal, or Saturn as the greater. It is a rare thing for a modern astrologer to introduce the place of an actual star into a horoscope

SATURN AND ASTROLOGY

the calculations all refer to the positions of the *Signs* of the Zodiac, which are purely imaginary divisions of the heavens; not to the *Constellations* of the Zodiac, which are the actual star-groups.

Until astronomers had determined the apparent orbits of the planets, and drawn up tables by which their apparent places could be predicted for some time in advance, it was impossible for astrologers to cast horoscopes of the present kind. All they could do was to divide up time amongst the deities supposed to preside over the various planets. To have simply given a planet to each day would have allowed the astrologer a very small scope in which to work for his prophecies; the ingenious idea of giving a planet to each hour as well, gave a wider range of possible combinations. There seems to have been deliberate spitefulness in the assignment of the most evil of the planetary divinities to the sacred day of the Jews—their sabbath. It should be noticed at the same time that, whilst the Jewish sabbath coincides with the astrological "Saturn's Day," that particular day is the seventh day of the Jewish week, but the first of the astrological. For the very nature of the reckoning by which the astrologers allotted the planets to the days of the week, implies, as shown in the extract quoted from Proctor, that they began with Saturn and worked downwards from the "highest planet" —as they called it—to the "lowest." This detail of itself should have sufficed to have demonstrated to Proctor, or any other astronomer, that the astrological week had been foisted upon the already existing week of the Jews.

Before astrology took its present mathematical form,

astrologers used as their material for prediction the stars or constellations which happened to be rising or setting at the time selected, or were upon the same meridian, or had the same longitude, as such constellations. One of the earliest of these astrological writers was Zeuchros of Babylon, who lived about the time of the Christian era, some of whose writings have been preserved to us. From these it is clear that the astrologers found twelve signs or the zodiac did not give them enough play. They therefore introduced the "decans," that is to say the idea of thirty-six divinities — three to each month — borrowed from the Egyptian division of the year into thirty-six weeks (of ten days), each under the rule of a separate god. Of course this Egyptian year bore no fixed relation to the actual lunar months or solar year, nor therefore to the Jewish year, which was related to both. But even with this increase of material, the astrologers found the astronomical data insufficient for their fortune-telling purposes. Additional figures quite unrepresented in the heavens, were devised, and were drawn upon, as needed, to supplement the genuine constellations, and as it was impossible to recognize these additions in the sky, the predictions were made, not from observation of the heavens, but from observations on globes, often very inaccurate.

Earlier still we have astrological tablets from Assyria and Babylon, many of which show that they had nothing to do with any actual observation, and were simply invented to give completeness to the tables of omens. Thus an Assyrian tablet has been found upon which are given the significations of eclipses falling upon each day of the month

SATURN AND ASTROLOGY   143

Tammuz, right up to the middle of the month. It is amusing to read the naïve comment of a distinguished Assyriologist, that tablets such as these prove how careful and how long continued had been the observations upon which they were based. It was not recognized that no eclipses either of sun or moon could possibly occur on most of the dates given, and that they could never occur "in the north," which is one of the quarters indicated. They were no more founded on actual observation than the portent mentioned on another tablet, of a woman giving birth to a lion, which, after all, is not more impossible than that an eclipse should occur in the north on the second day of Tammuz. In all ages it has been the same; the astrologer has had nothing to do with science as such, even in its most primitive form; he has cared nothing for the actual appearance of the heavens upon which he pretended to base his predictions; an imaginary planet, an imaginary eclipse, an imaginary constellation were just as good for his fortune-telling as real ones. Such fortune-telling was forbidden to the Hebrews; necessarily forbidden, for astrology had no excuse unless the stars and planets were gods, or the vehicles and engines of gods. Further, all attempts to extort from spirits or from inanimate things a glimpse into the future was likewise forbidden them. They were to look to God, and to His revealed will alone for all such light.

"When they shall say unto you, Seek unto them that have familiar spirits, and unto wizards that peep, and that mutter: should not a people seek unto their God?"

The Hebrews were few in number, their kingdoms very

small compared with the great empires of Egypt, Assyria, or Babylon, but here, in this question of divination or fortune-telling, they stand on a plane far above any of the surrounding nations. There is just contempt in the picture drawn by Ezekiel of the king of Babylon, great though his military power might be—

"The king of Babylon stood at the parting of the way, at the head of the two ways, to use divination: he shook the arrows to and fro, he consulted the teraphim, he looked in the liver."

And Isaiah calls upon the city of Babylon—

"Stand now with thine enchantments, and with the multitude of thy sorceries, wherein thou hast laboured from thy youth; if so thou shalt be able to profit, if so be thou mayest prevail. Thou art wearied in the multitude of thy counsels: let now the astrologers, the stargazers, the monthly prognosticators stand up, and save thee from these things that shall come upon thee."

Isaiah knew the Lord to be He that "frustrateth the tokens of the liars and maketh diviners mad." And the word of the Lord to Israel through Jeremiah was—

"Thus saith the Lord. Learn not the way of the heathen, and be not dismayed at the signs of heaven; for the heathen are dismayed at them."

It is to our shame that even to-day, in spite of all our enlightenment and scientific advances, astrology still has a hold upon multitudes. Astrological almanacs and treatises are sold by the tens of thousands, and astrological superstitions are still current. "The star of the god Chiun" is not indeed openly worshipped; but Saturn is

SATURN AND ASTROLOGY

still looked upon as the planet bringing such diseases as "toothache, agues, and all that proceeds from cold, consumption, the spleen particularly, and the bones, rheumatic gouts, jaundice, dropsy, and all complaints arising from fear, apoplexies, etc."; and charms made of Saturn's metal, lead, are still worn upon Saturn's finger, in the belief that these will ward off the threatened evil; a tradition of the time when by so doing the wearers would have proclaimed themselves votaries of the god, and therefore under his protection.

Astrology is inevitably linked with heathenism, and both shut up spirit and mind against the knowledge of God Himself, which is religion; and against the knowledge of His works, which is science. And though a man may be religious without being scientific, or scientific without being religious, religion and science alike both rest on one and the same basis—the belief in "One God, Maker of heaven and earth."

That belief was the reason why Israel of old, so far as it was faithful to it, was free from the superstitions of astrology.

"It is no small honour for this nation to have been wise enough to see the inanity of this and all other forms of divination. . . . Of what other ancient civilized nation could as much be said?"[1]

[1] G. V. Schiaparelli, *Astronomy in the Old Testament*, p. 52.

By permission of the Autotype Co.   *74, New Oxford Street, London, W.C.*

ST. PAUL PREACHING AT ATHENS (*by Raphael*).

"As certain also of your own poets have said, For we are also His offspring."

BOOK II

THE CONSTELLATIONS

CHAPTER I

THE ORIGIN OF THE CONSTELLATIONS

THE age of Classical astronomy began with the labours of Eudoxus and others, about four centuries before the Christian Era, but there was an Earlier astronomy whose chief feature was the arrangement of the stars into constellations.

The best known of all such arrangements is that sometimes called the "Greek Sphere," because those constellations have been preserved to us by Greek astronomers and poets. The earliest complete catalogue of the stars, as thus arranged, that has come down to us was compiled by Claudius Ptolemy, the astronomer of Alexandria, and completed 137 A.D. In this catalogue, each star is described by its place in the supposed figure of the constellation, whilst its celestial latitude and longitude are added, so that we can see with considerable exactness how the astronomers of that time imagined the star figures. The earliest complete description of the constellations, apart from the places of the individual stars, is given in the poem of Aratus of Soli—*The Phenomena*, published about 270 B.C.

Were these constellations known to the Hebrews of old? We can answer this question without hesitation in the case of St. Paul. For in his sermon to the Athenians on Mars' Hill, he quotes from the opening verses of this constellation poem of Aratus:—

"God that made the world and all things therein, seeing that He is Lord of heaven and earth, dwelleth not in temples made with hands; neither is worshipped with men's hands, as though He needed anything, seeing He giveth to all life, and breath, and all things; and hath made of one blood all nations of men for to dwell on all the face of the earth, and hath determined the times before appointed, and the bounds of their habitation; that they should seek the Lord, if haply they might feel after Him, and find Him, though He be not far from every one of us: for in Him we live, and move, and have our being; as certain also of your own poets have said, For we are also His offspring."

The poem of Aratus begins thus:—

"To God above we dedicate our song;
To leave Him unadored, we never dare;
For He is present in each busy throng,
In every solemn gathering He is there.
The sea is His; and His each crowded port;
In every place our need of Him we feel;
FOR WE HIS OFFSPRING ARE."

Aratus, like St. Paul himself, was a native of Cilicia, and had been educated at Athens. His poem on the constellations came, in the opinion of the Greeks, next in honour to the poems of Homer, so that St. Paul's quotation from it appealed to his hearers with special force.

The constellations of Ptolemy's catalogue are forty-eight in number. Those of Aratus correspond to them in

THE ORIGIN OF THE CONSTELLATIONS

almost every particular, but one or two minor differences may be marked. According to Ptolemy, the constellations are divided into three sets:—twenty-one northern, twelve in the zodiac, and fifteen southern.

The northern constellations are—to use the names by which they are now familiar to us—1, *Ursa Minor*, the Little Bear; 2, *Ursa Major*, the Great Bear; 3, *Draco*, the Dragon; 4, *Cepheus*, the King; 5, *Boötes*, the Herdsman; 6, *Corona Borealis*, the Northern Crown; 7, *Hercules*, the Kneeler; 8, *Lyra*, the Lyre or Swooping Eagle; 9, *Cygnus*, the Bird; 10, *Cassiopeia*, the Throned Queen, or the Lady in the Chair; 11, *Perseus*; 12, *Auriga*, the Holder of the Reins; 13, *Ophiuchus*, the Serpent-holder; 14, *Serpens*, the Serpent; 15, *Sagitta*, the Arrow; 16, *Aquila*, the Soaring Eagle; 17, *Delphinus*, the Dolphin; 18, *Equuleus*, the Horse's Head; 19, *Pegasus*, the Winged Horse; 20, *Andromeda*, the Chained Woman; 21, *Triangulum*, the Triangle.

The zodiacal constellations are : 1, *Aries*, the Ram; 2, *Taurus*, the Bull; 3, *Gemini*, the Twins; 4, *Cancer*, the Crab; 5, *Leo*, the Lion; 6, *Virgo*, the Virgin; 7, *Libra*, the Scales,—also called the Claws, that is of the Scorpion; 8, *Scorpio*, the Scorpion; 9, *Sagittarius*, the Archer; 10, *Capricornus*, the Sea-goat, *i. e.* Goat-fish; 11, *Aquarius*, the Water-pourer; 12, *Pisces*, the Fishes.

The southern constellations are: 1, *Cetus*, the Sea-Monster; 2, *Orion*, the Giant; 3, *Eridanus*, the River; 4, *Lepus*, the Hare; 5, *Canis Major*, the Great Dog; 6, *Canis Minor*, the Little Dog; 7, *Argo*, the Ship and Rock; 8, *Hydra*, the Water-snake; 9, *Crater*, the Cup; 10, *Corvus*, the Raven; 11, *Centaurus*, the Centaur; 12, *Lupus*, the Beast; 13, *Ara*, the Altar; 14, *Corona Australis*, the Southern Crown; 15, *Piscis Australis*, the Southern Fish.

Aratus, living four hundred years earlier than Ptolemy, differs only from him in that he reckons the cluster of the Pleiades—counted by Ptolemy in Taurus—as a separate constellation, but he has no constellation of *Equuleus*. The total number of constellations was thus still forty-eight. Aratus further describes the Southern Crown, but gives it no name; and in the constellation of the Little Dog he only mentions one star, *Procyon*, the Dog's Forerunner. He also mentions that the two Bears were also known as two Wagons or Chariots.

Were these constellations, so familiar to us to-day, known before the time of Aratus, and if so, by whom were they devised, and when and where?

They were certainly known before the time of Aratus, for his poem was confessedly a versification of an account of them written by Eudoxus more than a hundred years previous. At a yet earlier date, Panyasis, uncle to the great historian Herodotus, incidentally discusses the name of one of the constellations, which must therefore have been known to him. Earlier still, Hesiod, in the second book of his *Works and Days*, refers to several:—

> "Orion and the Dog, each other nigh,
> Together mounted to the midnight sky,
> When in the rosy morn Arcturus shines,
> Then pluck the clusters from the parent vines.
> . . . .
> Next in the round do not to plough forget
> When the Seven Virgins and Orion set."

THE ORIGIN OF THE CONSTELLATIONS 153

Much the same constellations are referred to by Homer. Thus, in the fifth book of the *Odyssey*,—

> "And now, rejoicing in the prosperous gales,
> With beating heart Ulysses spreads his sails :
> Placed at the helm he sate, and marked the skies,
> Nor closed in sleep his ever-watchful eyes.
> There view'd the Pleiads and the Northern Team,
> And great Orion's more refulgent beam,
> To which around the axle of the sky
> The Bear, revolving, points his golden eye."

Thus it is clear that several of the constellations were perfectly familiar to the Greeks a thousand years before the Christian era; that is to say, about the time of Solomon.

We have other evidence that the constellations were known in early times. We often find on Greek coins, a bull, a ram, or a lion represented; these may well be references to some of the signs of the zodiac, but offer no conclusive evidence. But several of the constellation figures are very unusual in form; thus the Sea-goat has the head and fore-legs of a goat, but the hinder part of a fish; and the Archer has the head and shoulders of a man, but the body and legs of a horse. Pegasus, the horse with wings, not only shows this unnatural combination, but the constellation figure only gives part of the animal —the head, neck, wings, breast, and fore-legs. Now some of these characteristic figures are found on quite early Greek coins, and yet earlier on what are known as "boundary stones" from Babylonia. These are little square pillars, covered with inscriptions and sculptures, and record for the most part the gift, transfer, or sale of land. They are dated according to the year of the reigning king, so that a clear idea can be formed as to their age. A great many symbols, which appear to be astronomical, occur upon them; amongst these such very distinguishing shapes as the Archer, Sea-goat, and Scorpion (*see* p. 318). So that, just as we know from Homer and Hesiod that the principal constellations were known of old by the same names as those by which we know them to-day, we learn from Babylonian boundary stones that they were then known as having the same forms as we now ascribe to them. The date of the earliest boundary stones of the kind in our possession would show that the Babylonians knew of our constellations as far back as the twelfth century B.C., that is to say, whilst Israel was under the Judges.

We have direct evidence thus far back as to the existence of the constellations. But they are older than this, so much older that tradition as well as direct historical evidence fails us. The only earlier evidence open to us is that of the constellations themselves.

A modern celestial globe is covered over with figures from pole to pole, but the majority of these are of quite recent origin and belong to the Modern period of astronomy. They have been framed since the invention of the telescope, and since the progress of geographical discovery brought men to know the southern skies. If these modern constellations are cleared off, and only those of Aratus and Ptolemy suffered to remain, it becomes at once evident that the ancient astronomers were not acquainted with the entire heavens. For there is a large space in the south,

THE ORIGIN OF THE CONSTELLATIONS 155

left free from all the old constellations, and no explanation, why it should have been so left free, is so simple and satisfactory as the obvious one, that the ancient astronomers did not map out the stars in that region because they

THE ANCIENT CONSTELLATIONS SOUTH OF THE ECLIPTIC.

never saw them; those stars never rose above their horizon.

Thus at the present time the heavens for an observer in England are naturally divided into three parts, as shown in the accompanying diagram. In the north, round the

156   THE ASTRONOMY OF THE BIBLE

pole-star are a number of constellations that never set; they wheel unceasingly around the pole. On every fine night we can see the Great Bear, the Little Bear, the Dragon, Cepheus and Cassiopeia. But the stars in the larger portion of the sky have their risings and settings,

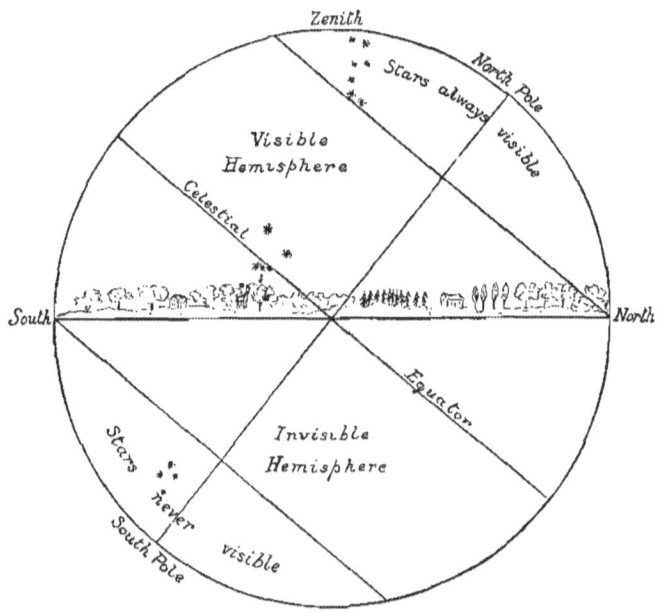

THE CELESTIAL SPHERE.

and the seasons in which they are visible or are withdrawn from sight. Thus we see Orion and the Pleiades and Sirius in the winter, not in the summer, but the Scorpion and Sagittarius in the summer. Similarly there is a third portion of the heavens which never comes within our range. We never see the Southern Cross, and hardly any

THE ORIGIN OF THE CONSTELLATIONS 157

star in the great constellation of the Ship, though these are very familiar to New Zealanders.

The outline of this unmapped region must therefore correspond roughly to the horizon of the place where the constellations were originally designed, or at least be roughly parallel to it, since we may well suppose that stars which only rose two or three degrees above that horizon might have been neglected.

From this we learn that the constellations were designed by people living not very far from the 40th parallel of north latitude, not further south than the 37th or 36th. This is important, as it shows that they did not originate in ancient Egypt or India, nor even in the city of Babylon, which is in latitude $32\frac{1}{2}°$.[1]

But this vacant space reveals another fact of even more importance. It gives us a hint as to the date when the constellations were designed.

An observer in north latitude 40° at the present time would be very far from seeing all the stars included in the forty-eight constellations. He would see nothing at all of the constellation of the Altar, and a good deal of that of the Centaur would be hidden from him.

. On the other hand, there are some bright constellations, such as the Phoenix and the Crane, unknown to the

---

[1] Delitzsch is, therefore, in error when he asserts that "when we divide the zodiac into twelve signs and style them the Ram, Bull, Twins, etc. . . . the Sumerian-Babylonian culture is still living and operating even at the present day" (*Babel and Bible*, p. 67). The constellations may have been originally designed by the *Akkadians*, but if so it was before they came down from their native highlands into the Mesopotamian valley.

158   THE ASTRONOMY OF THE BIBLE

ancients, which would come within his range of vision. This is due to what is known as "precession;" a slow movement of the axis upon which the earth rotates. In consequence of this, the pole of the heavens seems to trace out a circle amongst the stars which it takes 25,800 years to complete. It is therefore a matter of very simple calculation to find the position of the south pole of the heavens at any given date, past or future, and we find that the centre of the unmapped space was the south pole of the heavens something like 4,600 years ago, that is to say about 2,700 B.C.

It is, of course, not possible to fix either time or latitude very closely, since the limits of the unmapped space are a a little vague. But it is significant that if we take a celestial globe, arranged so as to represent the heavens for the time 2,700 B.C., and for north latitude 40°, we find several striking relations. First of all, the Great Dragon then linked together the north pole of the celestial equator, and the north pole of the ecliptic; it was as nearly as possible symmetrical with regard to the two; it occupied the very crown of the heavens. With the single exception of the Little Bear, which it nearly surrounds, the Dragon was the only constellation that never set. Next, the Watersnake (*see* diagram, p. 200) lay at this time right along the equator, extending over 105° of Right Ascension; or, to put it less technically, it took seven hours out of the twenty-four to cross the meridian. It covered nearly one-third of the equatorial belt. Thirdly, the intersection of the equator with one of the principal meridians of the sky was marked by the Serpent, which is carried by the

Serpent-holder in a very peculiar manner. The meridian at midnight at the time of the spring equinox is called a "colure,"—the "autumnal colure," because the sun crosses it in autumn. Now the Serpent was so arranged as to be shown writhing itself for some distance along the equator, and then struggling upwards, along the autumnal colure, marking the zenith with its head. The lower part of the autumnal colure was marked by the Scorpion, and the foot of the Serpent-holder pressed down the creature's head, just where the colure, the equator, and the ecliptic intersected (*see* diagram, p. 164).

It is scarcely conceivable that this fourfold arrangement, not suggested by any natural grouping of the stars, should have come about by accident; it must have been intentional. For some reason, the equator, the colure, the zenith and the poles were all marked out by these serpentine or draconic forms. The unmapped space gives us a clue only to the date and latitude of the designing of the most southerly constellations. We now see that a number of the northern hold positions which were specially significant under the same conditions, indicating that they were designed at about the same date. There is therefore little room for doubt that some time in the earlier half of the third millennium before our era, and somewhere between the 36th and 40th parallels of north latitude, the constellations were designed, substantially as we have them now, the serpent forms being intentionally placed in these positions of great astronomical importance.

It will have been noticed that Ptolemy makes the Ram the first constellation of the zodiac. It was so in his days, but it was the Bull that was the original leader, as we know from a variety of traditions; the sun at the spring equinox being in the centre of that constellation about 3000 B.C. At the time when the constellations were designed, the sun at the spring equinox was near Aldebaran, the brightest star of the Bull; at the summer solstice it was near Regulus, the brightest star of the Lion; at the autumnal equinox it was near Antares, the brightest star of the Scorpion; at the winter solstice it was near Fomalhaut, the brightest star in the neighbourhood of the Waterpourer. These four stars have come down to us with the name of the "Royal Stars," probably because they were so near to the four most important points in the apparent path of the sun amongst the stars. There is also a celebrated passage in the first of Virgil's *Georgics* which speaks of the white bull with golden horns that opens the year. So when the Mithraic religion adopted several of the constellation figures amongst its symbols, the Bull as standing for the spring equinox, the Lion for the summer solstice, were the two to which most prominence was given, and they are found thus used in Mithraic monuments as late as the second or third century A.D., long after the Ram had been recognized as the leading sign.

It is not possible to push back the origin of the constellations to an indefinite antiquity. They cannot at the very outside be more than 5000 years old; they must be considerably more than 4000. But during the whole of this millennium the sun at the spring equinox was in the constellation of the Bull. There is therefore no possible

THE ORIGIN OF THE CONSTELLATIONS

doubt that the Bull—and not the Twins nor the Ram—was the original leader of the zodiac.

The constellations, therefore, were designed long before the nation of Israel had its origin, indeed before Abraham left Ur of the Chaldees. The most probable date—2700 B.C.—would take us to a point a little before the Flood, if we accept the Hebrew chronology, a few centuries after the Flood, if we accept the Septuagint chronology. Just as the next great age of astronomical activity, which I have termed the Classical, began after the close of the canon of the Old Testament scriptures, so the constellation age began before the first books of those scriptures were compiled. Broadly speaking, it may be said that the knowledge of the constellation figures was the chief asset of astronomy in the centuries when the Old Testament was being written.

Seeing that the knowledge of these figures was preserved in Mesopotamia, the country from which Abraham came out, and that they were in existence long before his day, it is not unreasonable to suppose that both he and his descendants were acquainted with them, and that when he and they looked upward to the glories of the silent stars, and recalled the promise, "So shall thy seed be," they pictured round those glittering points of light much the same forms that we connect with them to-day.

CHAPTER II

GENESIS AND THE CONSTELLATIONS

As we have just shown, the constellations evidently were designed long before the earliest books of the Old Testament received their present form. But the first nine chapters of Genesis give the history of the world before any date that we can assign to the constellations, and are clearly derived from very early documents or traditions.

When the constellations are compared with those nine chapters, several correspondences appear between the two; remarkable, when it is borne in mind how few are the events that can be plainly set forth in a group of forty-eight figures on the one hand, and how condensed are the narratives of those nine chapters on the other.

Look at the six southern constellations (*see* pp. 164, 165) which were seen during the nights of spring in that distant time. The largest of these six is a great Ship resting on the southern horizon. Just above, a Raven is perched on the stretched-out body of a reptile. A figure of a Centaur appears to have just left the Ship, and is represented as offering up an animal on an Altar. The animal is now shown as a Wolf, but Aratus, our earliest authority, states that he did not know what kind of animal it was that

GENESIS AND THE CONSTELLATIONS 163

was being thus offered up. The cloud of smoke from the Altar is represented by the bright coiling wreaths of the Milky Way, and here in the midst of that cloud is set the Bow—the bow of Sagittarius, the Archer. Is it possible that this can be mere coincidence, or was it indeed intended as a memorial of the covenant which God made with Noah, and with his children for ever ?—" I do set My bow in the cloud, and it shall be for a token of a covenant between Me and the earth."

Close by this group was another, made up of five constellations. Towards the south, near midnight in spring, the observer in those ancient times saw the Scorpion. The figure of a man was standing upon that venomous beast, with his left foot pressed firmly down upon its head; but the scorpion's tail was curled up to sting him in the right heel. Ophiuchus, the Serpent-holder, the man treading on the Scorpion, derives his name from the Serpent which he holds in his hands and strangles; the Serpent that, as we have seen in the preceding chapter, marked the autumnal colure. The head of Ophiuchus reached nearly to the zenith, and there close to it was the head of another hero, so close that to complete the form of the two heads the same stars must be used to some extent twice over. Facing north, this second hero, now known to us as Hercules, but to Aratus simply as the " Kneeler," was seen kneeling with his foot on the head of the great northern Dragon. This great conflict between the man and the serpent, therefore, was presented in a twofold form. Looking south there was the picture of Ophiuchus trampling on the scorpion and strangling

164   THE ASTRONOMY OF THE BIBLE

the snake, yet wounded in the heel by the scorpion's sting; looking north, the corresponding picture of the kneeling figure of Hercules treading down the dragon's head. Here there seems an evident reference to the

THE MIDNIGHT CONSTELLATIONS OF SPRING, B.C. 2700.

word spoken by God to the serpent in the garden in Eden: "I will put enmity between thee and the woman, and between thy seed and her Seed; It shall bruise thy head, and thou shalt bruise His heel."

These two groups of star-figures seem therefore to point

GENESIS AND THE CONSTELLATIONS 165

to the two great promises made to mankind and recorded in the early chapters of Genesis; the Promise of the Deliverer, Who, " Seed of the woman," should bruise the serpent's head, and the promise of the " Bow set in the

THE MIDNIGHT CONSTELLATIONS OF WINTER, B.C. 2700.

cloud," the pledge that the world should not again be destroyed by a flood.

One or two other constellations appear, less distinctly, to refer to the first of these two promises. The Virgin, the woman of the zodiac, carries in her hand a bright star, the ear of corn, the seed; whilst, immediately under her, the great Water-snake, Hydra, is drawn out at enormous length, "going on its belly;" not writhing upwards like the Serpent, nor twined round the crown of the sky like the Dragon.

Yet again, the narrative in Genesis tells us that God "drove out the man" (*i. e.* Adam), "and He placed at the east of the garden of Eden the cherubim, and the flame of a sword which turned every way, to keep the way of the tree of life." No description is given of the form of the cherubim in that passage, but they are fully described by Ezekiel, who saw them in vision when he was by the river Chebar, as "the likeness of four living creatures." The same beings were also seen in vision by St. John, and are described by him in the Apocalypse as "four living creatures" (*Zōa*). "The first creature was like a lion, and the second creature like a calf, and the third creature had a face as of a man, and the fourth creature was like a flying eagle." Ezekiel gives a fuller and more complex description, but agreeing in its essential elements with that given by the Apostle, and, at the close of one of these descriptions, he adds, "This is the living creature that I saw under the God of Israel by the river of Chebar; and I knew that they were cherubim"—no doubt because as a priest he had been familiar with the cherubic forms as they were embroidered upon the curtains of the Temple, and carved upon its walls and doors.

The same four forms were seen amongst the constellation figures; not placed at random amongst them, but as far as possible in the four most important positions in the sky.

GENESIS AND THE CONSTELLATIONS 167

For the constellations were originally so designed that the sun at the time of the summer solstice was in the middle of the constellation *Leo*, the Lion; at the time of the spring equinox in the middle of *Taurus*, the Bull; and at the time of the winter solstice, in the middle of *Aquarius*, the Man bearing the waterpot. The fourth point, that held by the sun at the autumnal equinox, would appear to have been already assigned to the foot of the Serpent-holder as he crushes down the Scorpion's head; but a flying eagle, *Aquila*, is placed as near the equinoctial point as seems to have been consistent with the ample space that it was desired to give to the emblems of the great conflict between the Deliverer and the Serpent. Thus, as in the vision of Ezekiel, so in the constellation figures, the Lion, the Ox, the Man, and the Eagle, stood as the upholders of the firmament, as "the pillars of heaven." They looked down like watchers upon all creation; they seemed to guard the four quarters of the sky.

If we accept an old Jewish tradition, the constellations may likewise give us some hint of an event recorded in the tenth chapter of Genesis. For it has been supposed that the great stellar giant Orion is none other than "Nimrod, the mighty hunter before the Lord," and the founder of the Babylonian kingdom; identified by some Assyriologists with Merodach, the tutelary deity of Babylon: and by others with Gilgamesh, the tyrant of Erech, whose exploits have been preserved to us in the great epic now known by his name. Possibly both identifications may prove to be correct.

More than one third of the constellation figures thus appear to have a close connection with some of the chief incidents recorded in the first ten chapters of Genesis as having taken place in the earliest ages of the world's history. If we include the Hare and the two Dogs as adjuncts of Orion, and the Cup as well as the Raven with Hydra, then no fewer than twenty-two out of the forty-eight are directly or indirectly so connected. But the constellation figures only deal with a very few isolated incidents, and these are necessarily such as lend themselves to graphic representation. The points in common with the Genesis narrative are indeed striking, but the points of independence are no less striking. The majority of the constellation figures do not appear to refer to any incidents in Genesis; the majority of the incidents in the Genesis narrative find no record in the sky. Even in the treatment of incidents common to both there are differences, which make it impossible to suppose that either was directly derived from the other.

But it is clear that when the constellations were devised, —that is to say, roughly speaking, about 2,700 B.C.,—the promise of the Deliverer, the "Seed of the woman" who should bruise the serpent's head, was well known and highly valued; so highly valued that a large part of the sky was devoted to its commemoration and to that of the curse on the serpent. The story of the Flood was also known, and especially the covenant made with those who were saved in the ark, that the world should not again be destroyed by water, the token of which covenant was the "Bow set in the cloud." The fourfold cherubic forms were known, the keepers of the way of the tree of life, the symbols of the presence of God; and they were set in the four parts of the heaven, marking it out as the tabernacle which He spreadeth abroad, for He dwelleth between the cherubim.

CHAPTER III

THE STORY OF THE DELUGE

BESIDE the narrative of the Flood given to us in Genesis, and the pictorial representation of it preserved in the star figures, we have Deluge stories from many parts of the world. But in particular we have a very striking one from Babylonia. In the *Epic of Gilgamesh*, already alluded to, the eleventh tablet is devoted to an interview between the hero and Pir-napistim, the Babylonian Noah, who recounts to him how he and his family were saved at the time of the great flood.

This Babylonian story of the Deluge stands in quite a different relation from the Babylonian story of Creation in its bearing on the account given in Genesis. As we have already seen, the stories of Creation have practically nothing in common; the stories of the Deluge have many most striking points of resemblance, and may reasonably be supposed to have had a common origin.

Prof. Friedrich Delitzsch, in his celebrated lectures *Babel and Bible*, refers to this Babylonian Deluge story in the following terms:—

"The Babylonians divided their history into two great periods: the one before, the other after the Flood.

Babylon was in quite a peculiar sense the land of deluges. The alluvial lowlands along the course of all great rivers discharging into the sea are, of course, exposed to terrible floods of a special kind—cyclones and tornadoes accompanied by earthquakes and tremendous downpours of rain."

After referring to the great cyclone and tidal wave which wrecked the Sunderbunds at the mouths of the Ganges in 1876, when 215,000 persons met their death by drowning, Prof. Delitzsch goes on—

"It is the merit of the celebrated Viennese geologist, Eduard Suess, to have shown that there is an accurate description of such a cyclone, line for line, in the Babylonian Deluge story. . . . The whole story, precisely as it was written down, travelled to Canaan. But, owing to the new and entirely different local conditions, it was forgotten that the sea was the chief factor, and so we find in the Bible two accounts of the Deluge, which are not only scientifically impossible, but, furthermore, mutually contradictory—the one assigning to it a duration of 365 days, the other of $[40 + (3 \times 7)] = 61$ days. Science is indebted to Jean Astruc, that strictly orthodox Catholic physician of Louis XIV., for recognizing that two fundamentally different accounts of a deluge have been worked up into a single story in the Bible."[1]

The importance of the Babylonian Deluge story does not rest in anything intrinsic to itself, for there are many deluge stories preserved by other nations quite as interesting and as well told. It derives its importance from its points of resemblance to the Genesis story, and from the deduction that some have drawn from these that it was the original of that story—or rather of the two stories—that we find imperfectly recombined in Genesis.

[1] *Babel and Bible*, Johns' translation, pp. 42-46.

The suggestion of Jean Astruc that "two fundamentally different accounts of a deluge have been worked up into a single story in the Bible" has been generally accepted by those who have followed him in the minute analysis of the literary structure of Holy Scripture; and the names of the "Priestly Narrative" and of the "Jehovistic Narrative" have, for the sake of distinctness, been applied to them. The former is so called because the chapters in Exodus and the two following books, which treat with particular minuteness of the various ceremonial institutions of Israel, are considered to be by the same writer. The latter has received its name from the preference shown by the writer for the use, as the Divine name, of the word *Jehovah*,—so spelt when given in our English versions, but generally translated "the LORD."

There is a very close accord between different authorities as to the way in which Genesis, chapters vi.–ix., should be allotted to these two sources. The following is Dr. Driver's arrangement:—

	PRIESTLY NARRATIVE.			JEHOVISTIC NARRATIVE.	
	Chap.	Verse.		Chap.	Verse.
Genesis	vi.	9–22.	Genesis	vii.	1–5.
	vii.	6.			7–10.
		11.			12.
		13–16a.			16b.
		17a.			17b.
		18–21.			22–23.
		24.		viii.	2b–3a.
	viii.	1–2a.			6–12.
		3b–5.			13b.
		13a.			20–22.
		14–19.			
	ix.	1–17.			

THE STORY OF THE DELUGE 173

The Priestly narrative therefore tells us the cause of the Flood—that is to say, the corruption of mankind; describes the dimensions of the ark, and instructs Noah to bring "of every living thing of all flesh, two of every sort shalt thou bring into the ark, to keep them alive with thee; they shall be male and female." It further supplies the dates of the chief occurrences during the Flood, states that the waters prevailed above the tops of the mountains, that when the Flood diminished the ark rested upon the mountains of Ararat; and gives the account of Noah and his family going forth from the ark, and of the covenant which God made with them, of which the token was to be the bow seen in the cloud.

The most striking notes of the Jehovistic narrative are, —the incident of the sending out of the raven and the dove; the account of Noah's sacrifice; and the distinction made between clean beasts and beasts that are not clean— the command to Noah being, " Of every clean beast thou shalt take to thee by sevens, the male and his female: and of beasts that are not clean by two, the male and his female." The significant points of distinction between the two accounts are that the Priestly writer gives the description of the ark, the Flood prevailing above the mountains, the grounding on Mount Ararat, and the bow in the cloud; the Jehovistic gives the sending out of the raven and the dove, and the account of Noah's sacrifice, which involves the recognition of the distinction between the clean and unclean beasts and the more abundant provision of the former. He also lays emphasis on the Lord's "smelling a sweet savour" and promising never again to smite everything living, "for the imagination of man's heart is evil from his youth."

The chief features of the Babylonian story of the Deluge are as follows:—The God Ae spoke to Pir-napistim, the Babylonian Noah—

> "'Destroy the house, build a ship,
> Leave what thou hast, see to thy life.
> Destroy the hostile and save life.
> Take up the seed of life, all of it, into the midst of the ship.
> The ship which thou shalt make, even thou.
> Let its size be measured ;
> Let it agree as to its height and its length.'"

The description of the building of the ship seems to have been very minute, but the record is mutilated, and what remains is difficult to translate. As in the Priestly narrative, it is expressly mentioned that it was "pitched within and without."

The narrative proceeds in the words of Pir-napistim :—

"All I possessed, I collected it,
All I possessed I collected it, of silver ;
All I possessed I collected it, of gold ;
All I possessed I collected it, the seed of life, the whole.
I caused to go up into the midst of the ship,
All my family and relatives,
The beasts of the field, the animals of the field, the sons of the artificers—all of them I sent up.
The God Šamaš appointed the time—
Muir Kukki—'In the night I will cause the heavens to rain destruction,
Enter into the midst of the ship, and shut thy door.'
That time approached—
Muir Kukki— In the night the heavens rained destruction
I saw the appearance of the day :
I was afraid to look upon the day—
I entered into the midst of the ship, and shut my door

THE STORY OF THE DELUGE

At the appearance of dawn in the morning,
There arose from the foundation of heaven a dark cloud:
. . . . . .
The first day, the storm ? . . . .
Swiftly it swept, and . . . .
Like a battle against the people it sought.
Brother saw not brother.
The people were not to be recognized. In heaven
The gods feared the flood, and
They fled, they ascended to the heaven of Anu.
The gods kenneled like dogs, crouched down in the enclosures.
. . . . . .
The gods had crouched down, seated in lamentation,
Covered were their lips in the assemblies,
Six days and nights
The wind blew, the deluge and flood overwhelmed the land.
The seventh day, when it came, the storm ceased, the raging flood,
Which had contended like a whirlwind,
Quieted, the sea shrank back, and the evil wind and deluge ended.
I noticed the sea making a noise,
And all mankind had turned to corruption.
. . . . . .
I noted the regions, the shore of the sea,
For twelve measures the region arose.
The ship had stopped at the land of Nisir.
The mountain of Nisir seized the ship, and would not let it pass.
The first day and the second day the mountains of Nisir seized the
    ship, and would not let it pass.
. . . . . .
The seventh day, when it came
I sent forth a dove, and it left;
The dove went, it turned about,
But there was no resting-place, and it returned.
I sent forth a swallow, and it left,
The swallow went, it turned about,
But there was no resting-place, and it returned.
I sent forth a raven, and it left,
The raven went, the rushing of the waters it saw,
It ate, it waded, it croaked, it did not return.
I sent forth (the animals) to the four winds, I poured out a libation,
I made an offering on the peak of the mountain,

Seven and seven I set incense-vases there,
In their depths I poured cane, cedar, and rosewood (?).
The gods smelled a savour;
The gods smelled a sweet savour.
The gods gathered like flies over the sacrificer.
Then the goddess Sîrtu, when she came,
Raised the great signets that Anu had made at her wish:
'These gods—by the lapis-stone of my neck—let me not forget;
These days let me remember, nor forget them for ever!
Let the gods come to the sacrifice,
But let not Bêl come to the sacrifice,
For he did not take counsel, and made a flood,
And consigned my people to destruction.'
Then Bêl, when he came,
Saw the ship. And Bêl stood still,
Filled with anger on account of the gods and the spirits of heaven
What, has a soul escaped?
Let not a man be saved from the destruction.'
Ninip opened his mouth and spake.
He said to the warrior Bêl:
'Who but Ae has done the thing?
And Ae knows every event.'
Ae opened his mouth and spake,
He said to the warrior Bêl:
"Thou sage of the gods, warrior,
Verily thou hast not taken counsel, and hast made a flood.
The sinner has committed his sin,
The evil-doer has committed his misdeed,
Be merciful—let him not be cut off—yield, let not perish.
Why hast thou made a flood?
Let the lion come, and let men diminish.
Why hast thou made a flood?
Let the hyena come, and let men diminish.
Why hast thou made a flood?
Let a famine happen, and let the land be (?)
Why hast thou made a flood?
Let Ura (pestilence) come, and let the land be (?)'"[1]

---

[1] T. G. Pinches, *The Old Testament in the Light of the Historical Records of Assyria and Babylonia*, pp. 102-107.

THE STORY OF THE DELUGE 177

Of the four records before us, we can only date one approximately. The constellations, as we have already seen, were mapped out some time in the third millennium before our era, probably not very far from 2700 B.C.

When was the Babylonian story written? Does it, itself, afford any evidence of date? It occurs in the eleventh tablet of the *Epic of Gilgamesh*, and the theory has been started that as Aquarius, a watery constellation, is now the eleventh sign of the zodiac, therefore we have in this epic of twelve tablets a series of solar myths founded upon the twelve signs of the zodiac, the eleventh giving us a legend of a flood to correspond to the stream of water which the man in Aquarius pours from his pitcher.

If this theory be accepted we can date the *Epic of Gilgamesh* with much certainty: it must be later, probably much later, than 700 B.C. For it cannot have been till about that time that the present arrangement of the zodiacal signs—that is to say with Aries as the first and Aquarius as the eleventh—can have been adopted. We have then to allow for the growth of a mythology with the twelve signs as its *motif*. Had this supposed series of zodiacal myths originated before 700 B.C., before Aries was adopted as the leading sign, then the Bull, Taurus, would have given rise to the myth of the first tablet and Aquarius to the tenth, not to the eleventh where we find the story of the flood.

Assyriologists do not assign so late a date to this poem, and it must be noted that the theory supposes, not merely that the tablet itself, but that the poem and the series of myths upon which it was based, were all later in conception than 700 B.C. One conclusive indication of its early date is given by the position in the pantheon of Ae and Bêl. Ae has not receded into comparative insignificance, nor has Bêl attained to that full supremacy which, as Merodach, he possesses in the Babylonian Creation story. We may therefore put on one side as an unsupported and unfortunate guess the suggestion that the *Epic of Gilgamesh* is the setting forth of a series of zodiacal myths.

Any legends, any mythology, any pantheon based upon the zodiac must necessarily be more recent than the zodiac; any system involving Aries as the first sign of the zodiac must be later than the adoption of Aries as the first sign, that is to say, later than 700 B.C. Systems arising before that date would inevitably be based upon Taurus as first constellation.

We cannot then, from astronomical relationships, fix the date of the Babylonian story of the Flood. Is it possible, however, to form any estimate of the comparative ages of the Babylonian legend and of the two narratives given in Genesis, or of either of these two? Does the Babylonian story connect itself with one of the Genesis narratives rather than the other?

The significant points in the Babylonian story are these:—the command to Pir-napistim to build a ship, with detailed directions; the great rise of the flood so that even the gods in the heaven of Anu feared it; the detailed dating of the duration of the flood; the stranding of the ship on the mountain of Nisir; the sending forth

THE STORY OF THE DELUGE

of the dove, the swallow, and their return; the sending forth of the raven, and its non-return; the sacrifice; the gods smelling its sweet savour; the vow of remembrance of the goddess by the lapis-stone necklace; the determination of the gods not to send a flood again upon the earth, since sin is inevitable from the sinner. To all these points we find parallels in the account as given in Genesis.

But it is in the Priestly narrative that we find the directions for the building of the ship; the great prevalence of the flood even to the height of the mountains; the stranding of the ship on a mountain; and the bow in the clouds as a covenant of remembrance—this last being perhaps paralleled in the Babylonian story by the mottled (blue and white) lapis necklace of the goddess which she swore by as a remembrancer. There is therefore manifest connection with the narrative told by the Priestly writer.

But it is in the Jehovistic narrative, on the other hand, that we find the sending forth of the raven, and its non-return; the sending forth of the dove, and its return; the sacrifice, and the sweet savour that was smelled of the Lord; and the determination of the Lord not to curse the earth any more for man's sake, nor smite any more every living thing, "for the imagination of man's heart is evil from his youth." There is, therefore, no less manifest connection with the narrative told by the Jehovistic writer.

But the narrative told by the writer of the Babylonian story is one single account; even if it were a combination of two separate traditions, they have been so completely fused that they cannot now be broken up so as to form two distinct narratives, each complete in itself.

"The whole story precisely as it was written down travelled to Canaan,"—so we are told. And there,—we are asked to believe,—two Hebrew writers of very different temperaments and schools of thought, each independently worked up a complete story of the Deluge from this Gilgamesh legend. They chose out different incidents, one selecting what the other rejected, and *vice versa*, so that their two accounts were "mutually contradictory." They agreed, however, in cleansing it from its polytheistic setting, and giving it a strictly monotheistic tone. Later, an "editor" put the two narratives together, with all their inconsistencies and contradictions, and interlocked them into one, which presents all the main features of the original Gilgamesh story except its polytheism. In other words, two Hebrew scribes each told in his own way a part of the account of the Deluge which he had derived from Babylon, and a third unwittingly so recombined them as to make them represent the Babylonian original!

The two accounts of the Deluge, supposed to be present in Genesis, therefore cannot be derived from the Gilgamesh epic, nor be later than it, seeing that what is still plainly separable in Genesis is inseparably fused in the epic.

On the other hand, can the Babylonian narrative be later than, and derived from, the Genesis account? Since so many of the same circumstances are represented in

THE STORY OF THE DELUGE 181

both, this is a more reasonable proposition, if we assume that the Babylonian narrator had the Genesis account as it now stands, and did not have to combine two separate statements. For surely if he had the separate Priestly and Jehovistic narratives we should now be able to decompose the Babylonian narrative just as easily as we do the one in Genesis. The Babylonian adapter of the Genesis story must have either been less astute than ourselves, and did not perceive that he had really two distinct (and "contradictory") narratives to deal with, or he did not consider this circumstance of the slightest importance, and had no objection to merging them inextricably into one continuous account.

It is therefore possible that the Babylonian account was derived from that in Genesis; but it is not probable. The main circumstances are the same in both, but the details, the presentment, the attitude of mind are very different. We can better explain the agreement in the general circumstances, and even in many of the details, by presuming that both are accounts—genuine traditions—of the same actual occurrence. The differences in detail, presentment, and attitude, are fully and sufficiently explained by supposing that we have traditions from two, if not three, witnesses of the event.

We have also the pictorial representation of the Flood given us in the constellations. What evidence do they supply?

Here the significant points are: the ship grounded upon a high rock; the raven above it, eating the flesh of a stretched-out reptile; a sacrifice offered up by a person, who has issued forth from the ship, upon an altar, whose smoke goes up in a cloud, in which a bow is set.

In this grouping of pictures we have two characteristic features of the Priestly narrative, in the ship grounded on a rock, and in the bow set in the cloud; we have also two characteristic features of the Jehovistic narrative, in the smoking altar of sacrifice, and in the carrion bird. There is therefore manifest connection between the constellation grouping and *both* the narratives given in Genesis.

But the constellational picture story is the only one of all these narratives that we can date. It must have been designed—as we have seen—about 2700 B.C.

The question again comes up for answer. Were the Genesis and Babylonian narratives, any or all of them, derived from the pictured story in the constellations; or, on the other hand, was this derived from any or all of them?

The constellations were mapped out near the north latitude of 40°, far to the north of Babylonia, so the pictured story cannot have come from thence. We do not know where the Genesis narratives were written, but if the Flood of the constellations was pictured from them, then they must have been already united into the account that is now presented to us in Genesis, very early in the third millennium before Christ.

Could the account in Genesis have been derived from the constellations? If it is a double account, most decidedly not; since the pictured story in the constellations is one, and presents impartially the characteristic features of *both* the narratives.

THE STORY OF THE DELUGE 183

And (as in comparing the Genesis and the Babylonian narratives) we see that though the main circumstances are the same—in so far as they lend themselves to pictorial representations—the details, the presentment, the attitude are different. In the Genesis narrative, the bow set in the cloud is a rainbow in a cloud of rain; in the constellation picture, the bow set in the cloud is the bow of an archer, and the cloud is the pillar of smoke from off the altar of sacrifice. In the narratives of Genesis and Babylonia, Noah and Pir-napistim are men : no hint is given anywhere that by their physical form or constitution they were marked off from other men ; in the storied picture, he who issues from the ship is a centaur: his upper part is the head and body of a man, his lower part is the body of a horse.

As before, there is no doubt that we can best explain the agreement in circumstance of all the narratives by presuming that they are independent accounts of the same historical occurrence. We can, at the same time, explain the differences in style and detail between the narratives by presuming that the originals were by men of different qualities of mind who each wrote as the occurrence most appealed to him. The Babylonian narrator laid hold of the promise that, though beast, or famine, or pestilence might diminish men, a flood should not again sweep away every living thing, and connected the promise with the signets—the lapis necklace of the goddess Sirtu that she touched as a remembrancer. The picturer of the constellations saw the pledge in the smoke of the sacrifice, in the spirit of the words of the

Lord as given by Asaph, " Gather My saints together unto me; those that have made a covenant with Me by sacrifice." The writer in Genesis saw the promise in the rain-cloud, for the rainbow can only appear with the shining of the sun. The writer in Genesis saw in Noah a righteous man, worthy to escape the flood of desolation that swept away the wickedness around; there is no explanation apparent, at least on the surface, as to why the designer of the constellations made him, who issued from the ship and offered the sacrifice, a centaur—one who partook of two natures.

The comparison of the Deluge narratives from Genesis, from the constellations, and from Babylonia, presents a clear issue. If all the accounts are independent, and if there are two accounts intermingled into one in Genesis, then the chief facts presented in both parts of that dual narrative must have been so intermingled at an earlier date than 2700 B.C. The editor who first united the two stories into one must have done his work before that date.

But if the accounts are not independent histories, and the narrative as we have it in Genesis is derived either in whole or in part from Babylonia or from the constellations—if, in short, the Genesis story came from a Babylonian or a stellar myth—then we cannot escape from this conclusion: that the narrative in Genesis is not, and never has been, two separable portions; that the scholars who have so divided it have been entirely in error. But we cannot so lightly put on one side the whole of the results which the learning and research of so many scholars have given us in the last century-and-a-half. We must therefore unhesitatingly reject the theory that the Genesis Deluge story owes anything either to star myth or to Babylonian mythology. And if the Genesis Deluge story is not so derived, certainly no other portion of Holy Scripture.

CHAPTER IV

THE TRIBES OF ISRAEL AND THE ZODIAC

THE earliest reference in Scripture to the constellations of the zodiac occurs in the course of the history of Joseph. In relating his second dream to his brethren he said—

"Behold, I have dreamed a dream more; and, behold, the sun and the moon, and the eleven stars made obeisance to me."

The word "*Kochab*" in the Hebrew means both "star" and "constellation." The significance, therefore, of the reference to the "eleven stars" is clear. Just as Joseph's eleven brethren were eleven out of the twelve sons of Jacob, so Joseph saw eleven constellations out of the twelve come and bow down to him. And the twelve constellations can only mean the twelve of the zodiac.

There can be no reasonable doubt that the zodiac in question was practically the same as we have now, the one transmitted to us through Aratus and Ptolemy. It had been designed quite a thousand years earlier than the days of Joseph; it was known in Mesopotamia from whence his ancestors had come; it was known in Egypt; that is to say it was known on both sides of Canaan.

There have been other zodiacs: thus the Chinese have one of their own: but we have no evidence of any zodiac, except the one transmitted to us by the Greeks, as having been at any time adopted in Canaan or the neighbouring countries.

There is no need to suppose that each of the brethren had a zodiacal figure already assigned to him as a kind of armorial bearing or device. The dream was appropriate, and perfectly intelligible to Jacob, to Joseph, and his brethren, without supposing that any such arrangement had then been made. It is quite true that there are Jewish traditions assigning a constellation to each of the tribes of Israel, but it does not appear that any such traditions can be distinctly traced to a great antiquity, and they are mostly somewhat indefinite. Josephus, for instance, makes a vague assertion about the twelve precious stones of the High Priest's breast-plate, each of which bore the name of one of the tribes, connecting them with the signs of the zodiac:—

"Now the names of all those sons of Jacob were engraven in these stones, whom we esteem the heads of our tribes, each stone having the honour of a name, in the order according to which they were born. . . . And for the twelve stones whether we understand by them the months, or whether we understand the like number of the signs of that circle which the Greeks call the Zodiac, we shall not be mistaken in their meaning." [1]

But whilst there is no sufficient evidence that each of the sons of Jacob had a zodiacal figure for his coat-of-arms, nor even that the tribes deriving their names from

[1] Josephus, *Antiquities of the Jews*, III. vii. 5–7.

them were so furnished, there is strong and harmonious tradition as to the character of the devices borne on the standards carried by the four divisions of the host in the march through the wilderness. The four divisions, or camps, each contained three tribes, and were known by the name of the principal tribe in each. The camp of Judah was on the east, and the division of Judah led on the march. The camp of Reuben was on the south. The camp of Ephraim was on the west. The camp of Dan was on the north, and the division of Dan brought up the rear. And the traditional devices shown on the four standards were these:—For Judah, a lion; for Reuben, a man and a river; for Ephraim, a bull; for Dan, an eagle and a serpent.

In these four standards we cannot fail to see again the four cherubic forms of lion, man, ox and eagle; but in two cases an addition was made to the cherubic form, an addition recalling the constellation figure. For just as the crest of Reuben was not a man only, but a man and a river, so Aquarius is not a man only, but a man pouring out a stream of water. And as the crest of Dan was not an eagle only, but an eagle and a serpent, so the great group of constellations, clustering round the autumnal equinox, included not only the Eagle, but also the Scorpion and the Serpent (*see* diagram, p. 189).

There appears to be an obvious connection between these devices and the blessings pronounced by Jacob upon his sons, and by Moses upon the tribes; indeed, it would seem probable that it was the former that largely determined the choice of the devices adopted by the four great divisions of the host in the wilderness.

TRIBES OF ISRAEL AND THE ZODIAC 189

The blessing pronounced by Jacob on Judah runs, "Judah is a lion's whelp: from the prey, my son, thou art gone up: he stooped down, he couched as a lion, and as an old lion; who shall rouse him up?" "The Lion of the tribe of Judah" is the title given to our Lord Himself in the Apocalypse of St. John.

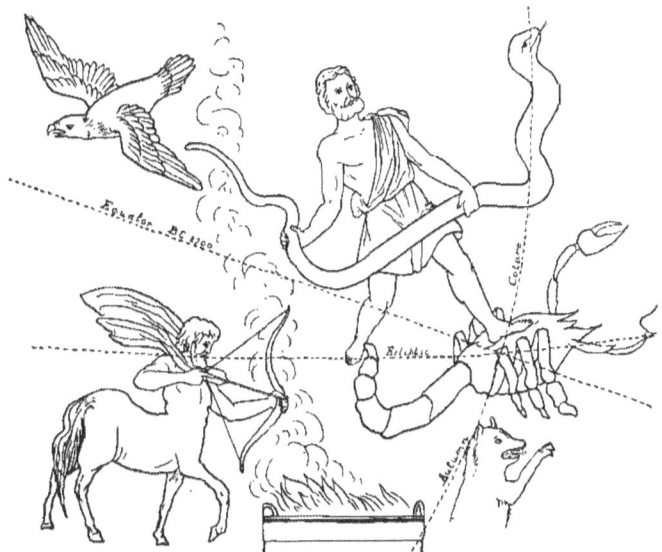

OPHIUCHUS AND THE NEIGHBOURING CONSTELLATIONS.

The blessing pronounced upon Joseph by Moses bears as emphatic a reference to the bull. "The firstling of his bullock, majesty is his; and his horns are the horns of the wild-ox."

Jacob's blessing upon Joseph does not show any reference to the ox or bull in our Authorized Version. But in our Revised Version Jacob says of Simeon and Levi—

"In their anger they slew a man,
And in their self-will they houghed an ox."

The first line appears to refer to the massacre of the Shechemites; the second is interpreted by the Jerusalem Targum, "In their wilfulness they sold Joseph their brother, who is likened to an ox." And in the blessing of Joseph it is said that his "branches (*margin*, daughters), run over the wall." Some translators have rendered this, "The daughters walk upon the bull," "wall" and "bull" being only distinguishable in the original by a slight difference in the pointing.

Of Reuben, his father said, "Unstable as water, thou shalt not excel;" and of Dan, "Dan shall be a serpent by the way, an adder in the path, that biteth the horse heels, so that his rider shall fall backward."

These two last prophecies supply the "water" and the "serpent," which, added to the "man" and "eagle" of the cherubic forms, are needed to complete the traditional standards, and are needed also to make them conform more closely to the constellation figures.

No such correspondence can be traced between the eight remaining tribes and the eight remaining constellations. Different writers combine them in different ways, and the allusions to constellation figures in the blessings of those tribes are in most cases very doubtful and obscure, even if it can be supposed that any such allusions are present at all. The connection cannot be pushed safely beyond the four chief tribes, and the four cherubic forms as represented in the constellations of the four quarters of the sky.

These four standards, or rather, three of them, meet us again in a very interesting connection. When Israel reached the borders of Moab, Balak, the king of Moab, sent for a seer of great reputation, Balaam, the son of Beor, to "Come, curse me Jacob, and come, defy Israel." Balaam came, but instead of cursing Jacob, blessed the people in four prophecies, wherein he made, what would appear to be, distinct references to the standards of Judah, Joseph and Reuben.

> "Behold the people riseth up as a lioness,
> And as a lion doth he lift himself up."

Then again—

> "He couched, he lay down as a lion,
> And as a lioness; who shall rouse him up?"

And in two passages—

> "God bringeth him forth out of Egypt;
> He hath as it were the strength of the wild ox."

The wild ox and lion are obvious similes to use concerning a powerful and warlike people. These two similes are, therefore, not sufficient by themselves to prove that the tribal standards are being referred to. But the otherwise enigmatical verse—

> "Water shall flow from his buckets,"

appears more expressly as an allusion to the standard of Reuben, the "man with the river," Aquarius pouring water from his pitcher; and if one be a reference to a standard, the others may also well be.

192   THE ASTRONOMY OF THE BIBLE

It is surely something more than coincidence that Joseph, who by his father's favour and his own merit was made

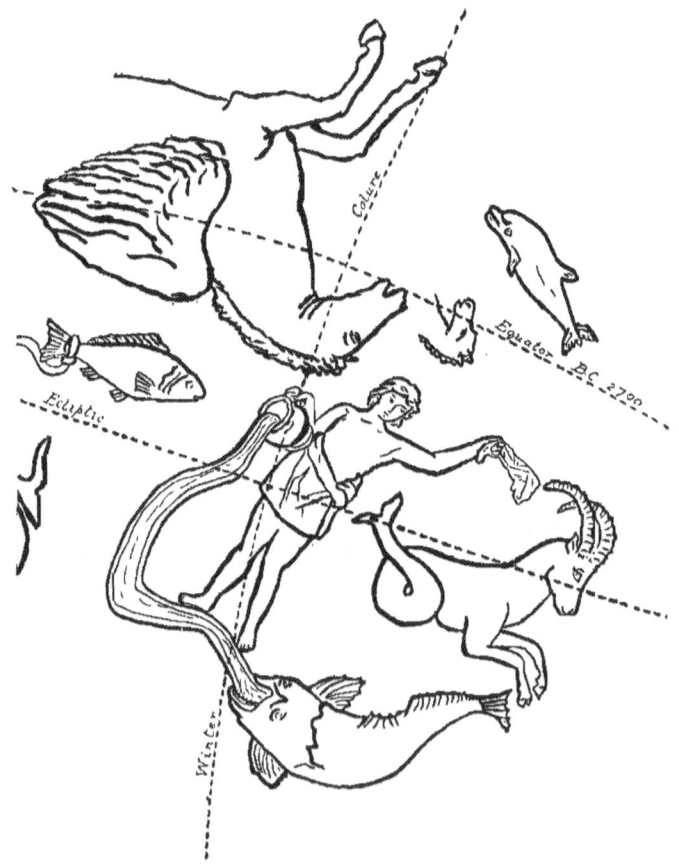

AQUARIUS AND THE NEIGHBOURING CONSTELLATIONS.

the leader of the twelve brethren, should be associated with the bull or wild ox, seeing that Taurus was the leader of the zodiac in those ages. It may also well be more than coincidence, that when Moses was in the mount and "the people gathered themselves unto Aaron, and said unto him, Up, make us gods, which shall go before us; for as for this Moses, the man that brought us up out of the land of Egypt, we wot not what is become of him," Aaron fashioned the golden earrings given him into the form of a molten calf; into the similitude, that is to say, of Taurus, then Prince of the Zodiac. If we turn to St. Stephen's reference to this occurrence, we find that he says—

"And they made a calf in those days, and offered sacrifice unto the idol, and rejoiced in the works of their own hands. Then God turned, and gave them up to worship the host of heaven."

In other words, their worship of the golden calf was star worship.

It has been often pointed out that this sin of the Israelites, deep as it was, was not in itself a breach of the first commandment—

"Thou shalt have no other gods before me."

It was a breach of the second—

"Thou shalt not make unto thee any graven image, or any likeness of anything that is in heaven above, or that is in the earth beneath, or that is in the water under the earth: thou shalt not bow down thyself to them nor serve them."

The Israelites did not conceive that they were abandoning the worship of Jehovah; they still considered themselves as worshipping the one true God. They were monotheists still, not polytheists. But they had taken the first false step that inevitably leads to polytheism; they had forgotten that they had seen "no manner of similitude on the day that the Lord spake unto" them "in Horeb out of the midst of the fire," and they had worshipped this golden calf as the similitude of God; they had "changed their glory into the similitude of an ox that eateth grass." And that was treason against Him; therefore St. Stephen said, "God turned, and gave them up to worship the host of heaven;" the one sin inevitably led to the other, indeed, involved it. In a later day, when Jeroboam, who had been appointed by Solomon ruler over all the charge of the house of Joseph, led the rebellion of the ten tribes against Rehoboam, king of Judah, he set up golden calves at Dan and Bethel, and said unto his people, "It is too much for you to go up to Jerusalem: behold thy gods, O Israel, which brought thee up out of the land of Egypt." There can be little doubt that, in this case, Jeroboam was not so much recalling the transgression in the wilderness—it was not an encouraging precedent—as he was adopting the well-known cognizance of the tribe of Joseph, that is to say, of the two tribes of Ephraim and Manasseh, which together made up the more important part of his kingdom, as the symbol of the presence of Jehovah.

The southern kingdom would naturally adopt the device of its predominant tribe, Judah, and it was as the undoubted cognizance of the kingdom of Judah that our Richard I., the Crusader, placed the Lion on his shield.

More definitely still, we find this one of the cherubic forms applied to set forth Christ Himself, as " The Root of David," Prince of the house of Judah—

" Behold, the Lion of the tribe of Juda, the Root of David, hath prevailed to open the book, and to loose the seven seals thereof."

CHAPTER V

LEVIATHAN

THERE are amongst the constellations four great draconic or serpent-like forms. Chief of these is the great dragon coiled round the pole of the ecliptic and the pole of the equator as the latter was observed some 4600 years ago. This is the dragon with which the Kneeler, *Hercules*, is fighting, and whose head he presses down with his foot. The second is the great watersnake, *Hydra*, which 4600 years ago stretched for 105° along the celestial equator of that day. Its head was directed towards the ascending node, that is to say the point where the ecliptic, the sun's apparent path, crosses the equator at the spring equinox; and its tail stretched nearly to the descending node, the point where the ecliptic again meets the equator at the autumn equinox. The third was the Serpent, the one held in the grip of the Serpent-holder. Its head erected itself just above the autumn equinox, and reached up as far as the zenith; its tail lay along the equator. The fourth of these draconic forms was the great Sea-monster, stretched out along the horizon, with a double river—*Eridanus*—proceeding from it, just below the spring equinox.

HERCULES AND DRACO.

None of these four figures was suggested by the natural grouping of the stars. Very few of the constellation-figures were so suggested, and these four in particular, as so high an authority as Prof. Schiaparelli expressly points out, were not amongst that few. Their positions show that they were designed some 4600 years ago, and that they have not been materially altered down to the present time. Though no forms or semblances of forms are there in the heavens, yet we still seem to see, as we look upwards, not merely the stars themselves, but the same snakes and dragons, first imagined so many ages ago as coiling amongst them.

The tradition of these serpentine forms and of their peculiar placing in the heavens was current among the Babylonians quite 1500 years after the constellations were devised. For the little "boundary stones" often display, amongst many other astronomical symbols, the coiled dragon round the top of the stone, the extended snake at its base (*see* p. 318), and at one or other corner the serpent bent into a right angle like that borne by the Serpent-holder—that is to say, the three out of the four serpentine forms that hold astronomically important positions in the sky.

The positions held by these three serpents or dragons have given rise to a significant set of astronomical terms. The Dragon marked the poles of both ecliptic and equator; the Watersnake marked the equator almost from node to node; the Serpent marked the equator at one of the nodes. The "Dragon's Head" and the "Dragon's Tail" therefore have been taken as astronomical symbols of the ascending and descending nodes of the sun's apparent path—the points where he seems to ascend above the equator in the spring, and to descend below it again in the autumn.

The moon's orbit likewise intersects the apparent path of the sun in two points, its two nodes; and the interval of time between its passage through one of these nodes and its return to that same node again is called a Draconic month, a month of the Dragon. The same symbols are applied by analogy to the moon's nodes.

Indeed the "Dragon's Head," ☊, is the general sign for the ascending node of any orbit, whether of moon, planet or comet, and the "Dragon's Tail," ☋, for the descending node. We not only use these signs in astronomical works to-day, but the latter sign frequently occurs, figured exactly as we figure it now, on Babylonian boundary stones 3000 years old.

But an eclipse either of the sun or of the moon can only take place when the latter is near one of its two nodes—is in the "Dragon's Head" or in the "Dragon's Tail." This relation might be briefly expressed by saying that the Dragon—that is of the nodes—causes the eclipse. Hence the numerous myths, found in so many nations, which relate how "a dragon devours the sun (or moon)" at the time of an eclipse.

The dragon of eclipse finds its way into Hindoo mythology in a form which shows clearly that the myth arose from a misunderstanding of the constellations. The equatorial Water-snake, stretching from one node nearly to the other, has resting upon it, *Crater*, the Cup. Combining this with the expression for the two nodes, the

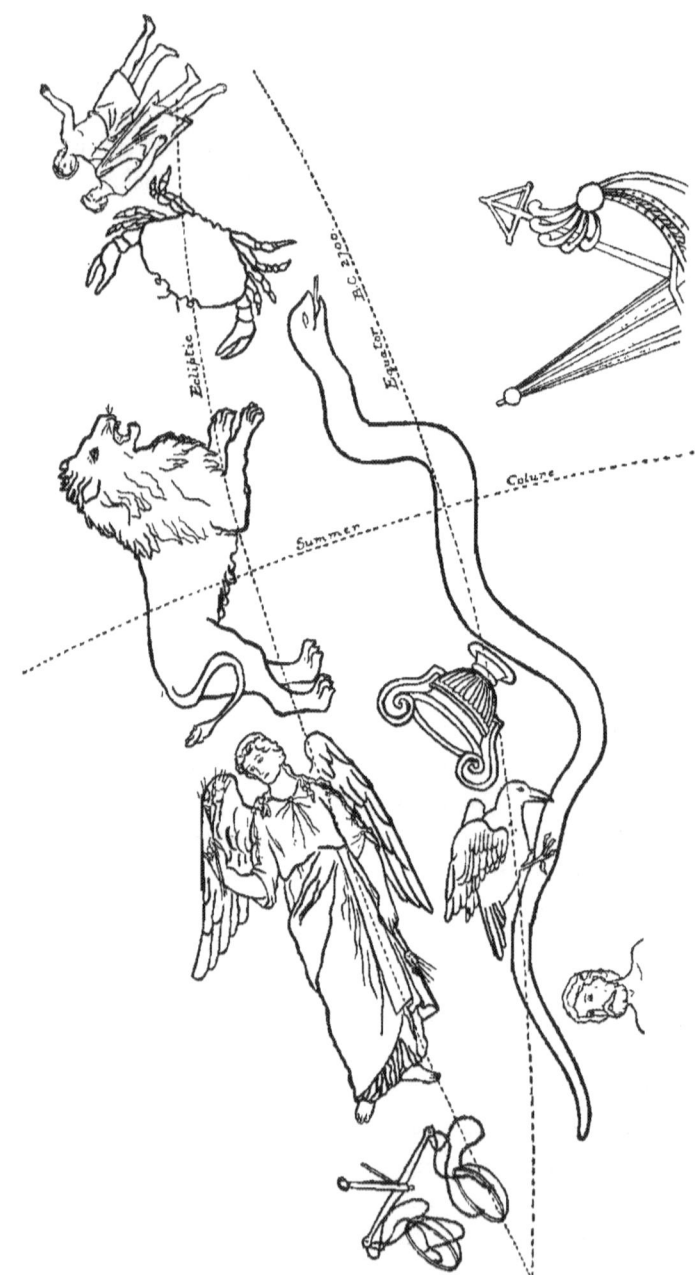

HYDRA AND THE NEIGHBOURING CONSTELLATIONS.

LEVIATHAN

Hindu myth has taken the following form. The gods churned the surface of the sea to make the Amrita Cup, the cup of the water of life. "And while the gods were drinking that nectar after which they had so much hankered, a Danava, named Rahu, was drinking it in the guise of a god. And when the nectar had only reached Rahu's throat, the sun and the moon discovered him, and communicated the fact to the gods." Rahu's head was at once cut off, but, as the nectar had reached thus far, it was immortal, and rose to the sky. "From that time hath arisen a long-standing quarrel between Rahu's head and the sun and moon," and the head swallows them from time to time, causing eclipses. Rahu's head marks the ascending, Ketu, the tail, the descending node.

This myth is very instructive. Before it could have arisen, not only must the constellations have been mapped out, and the equator and ecliptic both recognized, but the inclination of the moon's orbit to that of the sun must also have been recognized, together with the fact that it was only when the moon was near its node that the eclipses, either of the sun or moon, could take place. In other words, the cause of eclipses must have been at one time understood, but that knowledge must have been afterwards lost. We have seen already, in the chapter on "The Deep," that the Hebrew idea of *tehōm* could not possibly have been derived from the Babylonian myth of *Tiamat*, since the knowledge of the natural object must precede the myth founded upon it. If, therefore, Gen. i. and the Babylonian story of Creation be connected, the one as original, the other as derived from that original, it is the

Babylonian story that has been borrowed from the Hebrew, and it has been degraded in the borrowing.

So in this case, the myth of the Dragon, whose head and tail cause eclipses, must have been derived from a corruption and misunderstanding of a very early astronomical achievement. The myth is evidence of knowledge lost, of science on the down-grade.

Some may object that the myth may have brought about the conception of the draconic constellations. A very little reflection will show that such a thing was impossible. If the superstition that an eclipse is caused by an invisible dragon swallowing the sun or moon had really been the origin of the constellational dragons, they would certainly have all been put in the zodiac, the only region of the sky where sun or moon can be found; not outside it, where neither can ever come, and in consequence where no eclipse can take place. Nor could such a superstition have led on to the discoveries abovementioned: that the moon caused eclipses of the sun, the earth those of the moon; that the moon's orbit was inclined to the ecliptic, and that eclipses took place only near the nodes. The idea of an unseen spiritual agent being at work would prevent any search for a physical explanation, since polytheism is necessarily opposed to science.

There is a word used in Scripture to denote a reptilian monster, which appears in one instance at least to refer to this dragon of eclipse, and so to be used in an astronomical sense. Job, in his first outburst of grief cursed the day in which he was born, and cried—

LEVIATHAN

> "Let them curse it that curse the day,
> Who are ready (*margin*, skilful) to rouse up Leviathan.
> Let the stars of the twilight thereof be dark
> Let it look for light, but have none;
> Neither let it behold the eyelids of the morning."

"*Leviathan*" denotes an animal wreathed, gathering itself in coils: hence a serpent, or some great reptile. The description in Job xli. is evidently that of a mighty crocodile, though in Psalm civ. leviathan is said to play in "the great and wide sea," which has raised a difficulty as to its identification in the minds of some commentators. In the present passage it is supposed to mean one of the stellar dragons, and hence the mythical dragon of eclipse. Job desired that the day of his birth should have been cursed by the magicians, so that it had been a day of complete and entire eclipse, not even the stars that preceded its dawn being allowed to shine.

The astronomical use of the word *leviathan* here renders it possible that there may be in Isa. xxvii. an allusion—quite secondary and indirect however—to the chief stellar dragons.

> "In that day the Lord with His sore and great and strong sword shall punish leviathan the piercing serpent, even leviathan that crooked serpent; and He shall slay the dragon that is in the sea."

The marginal reading gives us instead of "piercing," "crossing like a bar"; a most descriptive epithet for the long-drawn-out constellation of *Hydra*, the Water-snake, which stretched itself for one hundred and five degrees along the primitive equator, and "crossed" the meridian "like a bar" for seven hours out of every twenty-four. "The crooked serpent" would denote the dragon coiled around the poles, whilst "the dragon which is in the sea" would naturally refer to *Cetus*, the Sea-monster. The prophecy would mean then, that "in that day" the Lord will destroy all the powers of evil which have, as it were, laid hold of the chief places, even in the heavens.

In one passage "the crooked serpent," here used as a synonym of *leviathan*, distinctly points to the dragon of the constellations. In Job's last answer to Bildad the Shuhite, he says—

"He divideth the sea with His power,
And by His understanding He smiteth through the proud.
(R.V. *Rahab*.)
By His spirit He hath garnished the heavens;
His hand hath formed the crooked serpent."

The passage gives a good example of the parallelism of Hebrew poetry; the repetition of the several terms of a statement, term by term, in a slightly modified sense; a rhyme, if the expression may be used, not of sound, but of signification.

Thus in the four verses just quoted, we have three terms in each—agent, action, object;—each appears in the first statement, each appears likewise in the second. The third statement, in like manner, has its three terms repeated in a varied form in the fourth.

Thus—

His power = His understanding.
Divideth = Smiteth through.
The sea = *Rahab* (the proud).

And—

His spirit	=	His hand.
Hath garnished	=	Hath formed.
The heavens	=	The crooked serpent.

There can be no doubt as to the significance of the two parallels. In the first, dividing the sea, *i.e.* the Red Sea, is the correlative of smiting through *Rahab*, "the proud one," the name often applied to Egypt, as in Isa. xxx. 7: "For Egypt helpeth in vain, and to no purpose: therefore have I called her Rahab that sitteth still." In the second, "adorning the heavens" is the correlative of "forming the crooked serpent." The great constellation of the writhing dragon, emphatically a "crooked serpent," placed at the very crown of the heavens, is set for all the constellations of the sky.

There are several references to *Rahab*, as "the dragon which is in the sea," all clearly referring to the kingdom of Egypt, personified as one of her own crocodiles lying-in-wait in her own river, the Nile, or transferred, by a figure of speech, to the Red Sea, which formed her eastern border. Thus in chapter li. Isaiah apostrophizes "the arm of the Lord."

"Art Thou not It that cut Rahab in pieces,
 That pierced the dragon?
Art Thou not It that dried up the sea,
 The waters of the great deep;
That made the depths of the sea a way for the redeemed to pass over?"

And in Psalm lxxxix. we have—

"Thou rulest the raging of the sea;
 When the waves thereof arise Thou stillest them.
Thou hast broken Rahab in pieces as one that is slain,
 Thou hast scattered Thine enemies with Thy strong arm."

So the prophet Ezekiel is directed—

"Son of man, take up a lamentation for Pharaoh, king of Egypt, and say unto him, thou wast likened unto a young lion of the nations: yet art thou as a dragon in the seas."

In all these passages it is only in an indirect and secondary sense that we can see any constellational references in the various descriptions of "the dragon that is in the sea." It is the crocodile of Egypt that is intended; Egypt the great oppressor of Israel, and one of the great powers of evil, standing as a representative of them all. The serpent or dragon forms in the constellations also represented the powers of evil; especially the great enemy of God and man, "the dragon, that old serpent, which is the Devil, and Satan." So there is some amount of appropriateness to the watery dragons of the sky—*Hydra* and *Cetus*—in these descriptions of *Rahab*, the dragon of Egypt, without there being any direct reference. Thus it is said of the Egyptian "dragon in the seas," I have given thee for meat to the beasts of the earth, and to the fowls of the heaven;" and again, "I will cause all the fowls of the heaven to settle upon thee," just as *Corvus*, the Raven, is shown as having settled upon *Hydra*, the Water-snake, and is devouring its flesh. Again, Pharaoh, the Egyptian dragon, says, "My river is mine own, and I have made it for myself;" just as *Cetus*, the Sea-monster, is represented as pouring forth *Eridanus*, the river, from its mouth.

But a clear and direct allusion to this last grouping of the constellations occurs in the Apocalypse. In the

LEVIATHAN 207

twelfth chapter, the proud oppressor dragon from the sea is shown us again with much fulness of detail. There the Apostle describes his vision of a woman, who evidently

ANDROMEDA AND CETUS.

represents the people of God, being persecuted by a dragon. There is still a reminiscence of the deliverance of Israel in the Exodus from Egypt, for " the woman *fled*.

into the wilderness, where she hath a place prepared of God, that there they may nourish her a thousand two hundred and threescore days." And the vision goes on:—

" And the serpent cast out of his mouth, after the woman water as a river, that he might cause her to be carried away by the stream. And the earth helped the woman, and the earth opened her mouth, and swallowed up the river which the dragon cast out of his mouth."

This appears to be precisely the action which is presented to us in the three constellations of *Andromeda, Cetus*, and *Eridanus*. Andromeda is always shown as a woman in distress, and the Sea-monster, though placed far from her in the sky, has always been understood to be her persecutor. Thus Aratus writes—

> "Andromeda, though far away she flies,
> Dreads the Sea-monster, low in southern skies."

The latter, baffled in his pursuit of his victim, has cast the river, *Eridanus*, out of his mouth, which, flowing down below the southern horizon, is apparently swallowed up by the earth.

It need occasion no surprise that we should find imagery used by St. John in his prophecy already set forth in the constellations nearly 3,000 years before he wrote. Just as, in this same book, St. John repeated Daniel's vision of the fourth beast, and Ezekiel's vision of the living creatures, as he used the well-known details of the Jewish Temple, the candlesticks, the laver, the altar of incense, so he used a group of stellar figures perfectly well known at the time when he wrote. In so doing the beloved

LEVIATHAN

disciple only followed the example which his Master had already set him. For the imagery in the parables of our Lord is always drawn from scenes and objects known and familiar to all men.

In two instances in which *leviathan* is mentioned, a further expression is used which has a distinct astronomical bearing. In the passage already quoted, where Job curses the day of his birth, he desires that it may not "behold the eyelids of the morning." And in the grand description of *leviathan*, the crocodile, in chapter xli., we have—

"His neesings flash forth light,
And his eyes are like the eyelids of the morning."

Canon Driver considers this as an "allusion, probably to the reddish eyes of the crocodile, which are said to appear gleaming through the water before the head comes to the surface." This is because of the position of the eyes on the animal's head, not because they have any peculiar brilliancy.

"It is an idea exclusively Egyptian, and is another link in the chain of evidence which connects the author of the poem with Egypt. The crocodile's head is so formed that its highest points are the eyes; and when it rises obliquely to the surface the eyes are the first part of the whole animal to emerge. The Egyptians observing this, compared it to the sun rising out of the sea, and made it the hieroglyphic representative of the idea of sunrise. Thus Horus Apollo says: When the Egyptians represent the sunrise, they paint the eye of the crocodile, because it is first seen as that animal emerges from the water."[1]

[1] P. H. Gosse, in the *Imperial Bible-Dictionary*.

In this likening of the eyes of the crocodile to the eyelids of the morning, we have the comparison of one natural object with another. Such comparison, when used in one way and for one purpose, is the essence of poetry; when used in another way and for another purpose, is the essence of science. Both poetry and science are opposed to myth, which is the confusion of natural with imaginary objects, the mistaking the one for the other.

Thus it is poetry when the Psalmist speaks of the sun "as a bridegroom coming out of his chamber"; for there is no confusion in his thought between the two natural objects. The sun is like the bridegroom in the glory of his appearance. The Psalmist does not ascribe to him a bride and children.

It is science when the astronomer compares the spectrum of the sun with the spectra of various metals in the laboratory. He is comparing natural object with natural object, and is enabled to draw conclusions as to the elements composing the sun, and the condition in which they there exist.

But it is myth when the Babylonian represents Bel or Merodach as the solar deity, destroying Tiamat, the dragon of darkness, for there is confusion in the thought. The imaginary god is sometimes given solar, sometimes human, sometimes superhuman characteristics. There is no actuality in much of what is asserted as to the sun or as to the wholly imaginary being associated with it. The mocking words of Elijah to the priests of Baal were justified by the intellectual confusion of their ideas, as well as by the spiritual degradation of their idolatry.

"Cry aloud: for he is a god; either he is talking, or he is pursuing, or he is in a journey, or peradventure he sleepeth, and must be awakened."

Such nature-myths are not indications of the healthy mental development of a primitive people; they are the clear signs of a pathological condition, the symptoms of intellectual disease.

It is well to bear in mind this distinction, this opposition between poetry and myth, for ignoring it has led to not a little misconception as to the occurrence of myth in Scripture, especially in connection with the names associated with the crocodile. Thus it has been broadly asserted that "the original mythical signification of the monsters *tehôm, livyāthān, tannim, rahâb*, is unmistakably evident."

Of these names the first signifies the world of waters; the second and third real aquatic animals; and the last, "the proud one," is simply an epithet of Egypt, applied to the crocodile as the representation of the kingdom. There is no more myth in setting forth Egypt by the crocodile or leviathan than in setting forth Great Britain by the lion, or Russia by the bear.

The Hebrews in setting forth their enemies by crocodile and other ferocious reptiles were not describing any imaginary monsters of the primæval chaos, but real oppressors. The Egyptian, with his "house of bondage," the Assyrian, "which smote with a rod," the Chaldean who made havoc of Israel altogether, were not dreams. And in beseeching God to deliver them from their latest oppressor the Hebrews naturally recalled, not some idle tale of the fabulous achievements of Babylonian deities, but the actual deliverance God had wrought for them at the Red Sea. There the Egyptian crocodile had been made "meat to the people inhabiting the wilderness" when the corpses of Pharaoh's bodyguard, cast up on the shore, supplied the children of Israel with the weapons and armour of which they stood in need. So in the day of their utter distress they could still cry in faith and hope—

> "Yet God is my King of old,
> Working salvation in the midst of the earth.
> Thou didst divide the sea by Thy strength:
> Thou brakest the heads of the dragons in the waters.
> Thou brakest the heads of leviathan in pieces,
> And gavest him to be meat to the people inhabiting the wilderness.
> Thou didst cleave the fountain and the flood:
> Thou driedst up mighty rivers.
> The day is Thine, the night also is Thine:
> Thou hast prepared the light and the sun.
> Thou hast set all the borders of the earth:
> Thou hast made summer and winter."

CHAPTER VI

THE PLEIADES

THE translators of the Bible, from time to time, find themselves in a difficulty as to the correct rendering of certain words in the original. This is especially the case with the names of plants and animals. Some sort of clue may be given by the context, as, for instance, if the region is mentioned in which a certain plant is found, or the use that is made of it; or, in the case of an animal, whether it is "clean" or "unclean," what are its habits, and with what other animals it is associated. But in the case of the few Scripture references to special groups of stars, we have no such help. We are in the position in which Macaulay's New Zealander might be, if, long after the English nation had been dispersed, and its language had ceased to be spoken amongst men, he were to find a book in which the rivers "Thames," "Trent," "Tyne," and "Tweed" were mentioned by name, but without the slightest indication of their locality. His attempt to fit these names to particular rivers would be little more than a guess—a guess the accuracy of which he would have no means for testing.

This is somewhat our position with regard to the four

Hebrew names, *Kīmah, Kĕsīl, ʿAyish,* and *Mazzaroth*; yet in each case there are some slight indications which have given a clue to the compilers of our Revised Version, and have, in all probability, guided them correctly.

The constellations are not all equally attractive. A few have drawn the attention of all men, however otherwise inattentive. North-American Indians and Australian savages have equally noted the flashing brilliancy of Orion, and the compact little swarm of the Pleiades. All northern nations recognize the seven bright stars of the Great Bear, and they are known by a score of familiar names. They are the "Plough," or "Charles's Wain" of Northern Europe; the "Seven Plough Oxen" of ancient Rome; the "Bier and Mourners" of the Arabs; the "Chariot," or "Waggon," of the old Chaldeans; the "Big Dipper" of the prosaic New England farmer. These three groups are just the three which we find mentioned in the earliest poetry of Greece. So Homer writes, in the Fifth Book of the *Odyssey*, that Ulysses—

>"There view'd the Pleiads, and the Northern Team,
>And Great Orion's more refulgent beam,
>To which, around the axle of the sky,
>The Bear, revolving, points his golden eye."

It seems natural to conclude that these constellations, the most striking, or at all events the most universally recognized, would be those mentioned in the Bible.

The passages in which the Hebrew word *Kīmah*, is used are the following—

THE PLEIADES

(God) "maketh Arcturus, Orion, and Pleiades (*Kīmah*), and the chambers of the south" (Job ix. 9).

"Canst thou bind the sweet influences of Pleiades (*Kīmah*), or loose the bands of Orion?" (Job xxxviii. 31).

"Seek Him that maketh the seven stars (*Kīmah*) and Orion" (Amos v. 8).

In our Revised Version, *Kīmah* is rendered "Pleiades" in all three instances, and of course the translators of the Authorized Version meant the same group by the "seven stars" in their free rendering of the passage from Amos. The word *kīmah* signifies "a heap," or "a cluster," and would seem to be related to the Assyrian word *kimtu*, "family," from a root meaning to "tie," or "bind"; a family being a number of persons bound together by the very closest tie of relationship. If this be so we can have no doubt that our translators have rightly rendered the word. There is one cluster in the sky, and one alone, which appeals to the unaided sight as being distinctly and unmistakably a family of stars—the Pleiades.

The names '*Ash*, or '*Ayish*, *Kĕsīl*, and *Kīmah* are peculiar to the Hebrews, and are not, so far as we have any evidence at present, allied to names in use for any constellation amongst the Babylonians and Assyrians; they have, as yet, not been found on any cuneiform inscription. Amos, the herdsman of Tekoa, living in the eighth century B.C., two centuries before the Jews were carried into exile to Babylon, evidently knew well what the terms signified, and the writer of the Book of Job was no less aware of their signification. But the "Seventy," who translated the Hebrew Scriptures into Greek, were not at all clear as to the identification of these names of constellations; though they made their translation only two or three centuries after the Jews returned to Jerusalem under Ezra and Nehemiah, when oral tradition should have still supplied the meaning of such astronomical terms. Had these names been then known in Babylon, they could not have been unknown to the learned men of Alexandria in the second century before our era, since at that time there was a very direct scientific influence of the one city upon the other. This Hebrew astronomy was so far from being due to Babylonian influence and teaching, that, though known centuries before the exile, after the exile we find the knowledge of its technical terms was lost. On the other hand, *kīma* was the term used in all Syriac literature to denominate the Pleiades, and we accordingly find in the Peschitta, the ancient Syriac version of the Bible, made about the second century A.D., the term *kīma* retained throughout, but *kesil* and *'ayish* were reduced to their supposed Syriac equivalents.

Whatever uncertainty was felt as to the meaning of *kīmah* by the early translators, it is not now seriously disputed that the Pleiades is the group of stars in question.

The word *kīmah* means, as we have seen, "cluster" or "heap," so also the word *Pleiades*, which we use to-day, is probably derived from the Greek *Pleiones*, "many." Several Greek poets—Athenæus, Hesiod, Pindar, and Simonides—wrote the word *Peleiades*, i. e. "rock pigeons," considered as flying from the Hunter Orion; others made them the seven doves who carried ambrosia to the infant

Zeus. D'Arcy Thompson says, "The Pleiad is in many languages associated with bird-names, . . . and I am inclined to take the bird on the bull's back in coins of Eretria, Dicæa, and Thurii for the associated constellation of the Pleiad"[1]—the Pleiades being situated on the shoulder of Taurus the Bull.

The Hyades were situated on the head of the Bull, and in the Euphrates region these two little groups of stars were termed together, *Mas-tab-ba-gal-gal-la*, the Great Twins of the ecliptic, as Castor and Pollux were the Twins of the zodiac. In one tablet '*Îmina bi*, "the seven-fold one," and *Gut-dûa*, "the Bull-in-front," are mentioned side by side, thus agreeing well with their interpretation of "Pleiades and Hyades." The Semitic name for the Pleiades was also *Témennu*; and these groups of stars, worshipped as gods by the Babylonians, may possibly have been the *Gad* and *Meni*, " that troop," and " that number," referred to by the prophet Isaiah (lxv. 11).

On many Babylonian cylinder seals there are engraved seven small discs, in addition to other astronomical symbols. These seven small stellar discs are almost invariably arranged in the form : : : · or : : : · much as we should now-a-days plot the cluster of the Pleiades when mapping on a small scale the constellations round the Bull. It is evident that these seven little stellar discs do not mean the "seven planets," for in many cases the astronomical symbols which accompany them include both those of the sun and moon. It is most probable that they signify the Pleiades, or perhaps alternatively the Hyades.

[1] *Glossary of Greek Birds*, pp. 28, 29.

Possibly, reference is made to the worship of the Pleiades when the king of Assyria, in the seventh century B.C., brought men from Babylon and other regions to inhabit the depopulated cities of Samaria, "and the men of Babylon made Succoth-benoth." The Rabbis are said to have rendered this by the "booths of the Maidens," or the "tents of the Daughters,"—the Pleiades being the maidens in question.

Generally they are the Seven Sisters. Hesiod calls them the Seven Virgins, and the Virgin Stars. The names given to the individual stars are those of the seven daughters of Atlas and Pleione; thus Milton terms them the Seven Atlantic Sisters.

As we have seen (p. 189), the device associated expressly with Joseph is the Bull, and Jacob's blessing to his son has been sometimes rendered—

"Joseph is a fruitful bough, even a fruitful bough by a well; *the daughters walk upon the bull.*"

That is, " the Seven Sisters," the Pleiades, are on the shoulder of Taurus.

Aratus wrote of the number of the Pleiades—

"Seven paths aloft men say they take,
 Yet six alone are viewed by mortal eyes.
 From Zeus' abode no star unknown is lost,
 Since first from birth we heard, but thus the tale is told."

Euripides speaks of these "seven paths," and Eratosthenes calls them " the seven-starred Pleiad," although he describes one as " All-Invisible." There is a surprisingly universal tradition that they "were seven who now are

THE PLEIADES

six." We find it not only in ancient Greece and Italy, but also among the black fellows of Australia, the Malays of Borneo, and the negroes of the Gold Coast. There must be some reason to account for this widespread tradition. Some of the stars are known to be slightly variable, and one of the fainter stars in the cluster may have shone more brightly in olden time;—the gaseous

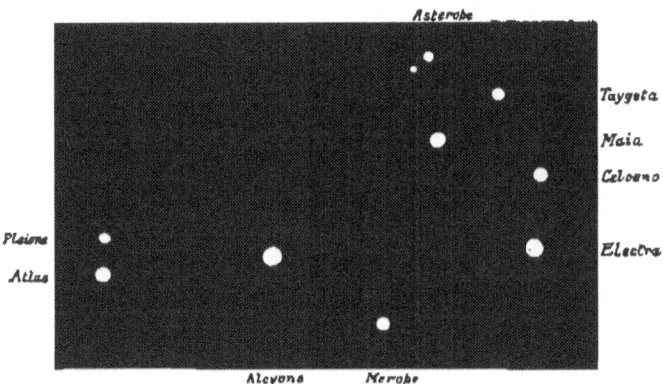

STARS OF THE PLEIADES.

spectrum of Pleione renders it credible that this star may once have had great brilliancy. Alcyone, now the brightest star in the cluster, was not mentioned by Ptolemy among the four brightest Pleiads of his day. The six now visible to ordinary sight are Alcyone, Electra, Atlas, Maia, Merope and Taygeta. Celoeno is the next in brightness, and the present candidate for the seventh place. By good sight, several more may be made out: thus Maestlin, the tutor of Kepler, mapped eleven before the invention of the telescope, and in our own day

Carrington and Denning have counted fourteen with the naked eye.

In clear mountain atmosphere more than seven would be seen by any keen-sighted observer. Usually six stars may be made out with the naked eye in both the Pleiades and the Hyades, or, if more than six, then several more; though with both groups the number of "seven" has always been associated.

In the New Testament we find the "Seven Stars" also mentioned. In the first chapter of the Revelation, the Apostle St. John says that he "saw seven golden candlesticks; and in the midst of the seven candlesticks one like unto the Son of Man, . . . and He had in His right hand seven stars." Later in the same chapter it is explained that "the seven stars are the angels of the seven churches; and the seven candlesticks which thou sawest are the seven churches." The seven stars in a single compact cluster thus stand for the Church in its many diversities and its essential unity.

This beautiful little constellation has become associated with a foolish fable. When it was first found that not only did the planets move round the sun in orbits, but that the sun itself also was travelling rapidly through space, a German astronomer, Mädler, hazarded the suggestion that the centre of the sun's motion lay in the Pleiades. It was soon evident that there was no sufficient ground for this suggestion, and that many clearly established facts were inconsistent with it. Nevertheless the idea caught hold of the popular mind, and it has acquired an amazing vogue. Non-astronomical writers have asserted that

Alcyone, the brightest Pleiad, is the centre of the entire universe; some have even been sufficiently irreverent to declare that it is the seat of heaven, the throne of God. A popular London divine, having noticed a bright ring round Alcyone on a photograph of the group, took that halo, which every photographer would at once recognize as a mere photographic defect, as a confirmation of this baseless fancy. Foolishness of this kind has nothing to support it in science or religion; it is an offence against both. We have no reason to regard the Pleiades as the centre of the universe, or as containing the attracting mass which draws our sun forward in its vast mysterious orbit.

R. H. Allen, in his survey of the literature of the Pleiades, mentions that "Drach surmised that their midnight culmination in the time of Moses, ten days after the autumnal equinox, may have fixed the Day of Atonement on the 10th of Tishri."[1] This is worth quoting as a sample of the unhappy astronomical guesses of commentators. Drach overlooked that his suggestion necessitated the assumption that in the time of Moses astronomers had already learned, first, to determine the actual equinox; next, to observe the culmination of stars on the meridian rather than their risings and settings; and, third and more important, to determine midnight by some artificial measurement of time. None of these can have been primitive operations; we have no knowledge that any of the three were in use in the time of Moses; certainly they were not suitable for a people on the march, like the Israelites in the wilderness. Above all, Drach

[1] R. H. Allen, *Star Names and their Meanings*, p. 401.

ignored in this suggestion the fact that the Jewish calendar was a lunar-solar one, and hence that the tenth day of the seventh month could not bear any fixed relation either to the autumnal equinox, or to the midnight culmination of the Pleiades; any more than our Easter Sunday is fixed to the spring equinox on March 22.

The Pleiades were often associated with the late autumn, as Aratus writes—

"Men mark them rising with Sol's setting light,
Forerunners of the winter's gloomy night,"

This is what is technically known as the "acronical rising" of the Pleiades, their rising at sunset; in contrast to their "heliacal rising," their rising just before daybreak, which ushered in the spring time. This acronical rising has led to the association of the group with the rainy season, and with floods. Thus Statius called the cluster "Pliadum nivosum sidus," and Valerius Flaccus distinctly used the word "Pliada" for the showers. Josephus says that during the siege of Jerusalem by Antiochus Epiphanes in 170 B.C., the besieged wanted for water until relieved "by a large shower of rain which fell at the setting of the Pleiades." R. H. Allen, in his *Star-Names and their Meanings*, states that the Pleiades "are intimately connected with traditions of the flood found among so many and widely separated nations, and especially in the Deluge-myth of Chaldæa," but he does not cite authorities or instances.

The Talmud gives a curious legend connecting the Pleiades with the Flood :—

" When the Holy One, blessed be He! wished to bring

THE PLEIADES

the Deluge upon the world, He took two stars out of Pleiades, and thus let the Deluge loose. And when He wished to arrest it, He took two stars out of Arcturus and stopped it."[1]

It would seem from this that the Rabbis connected the number of visible stars with the number of the family in the Ark—with the "few, that is, eight souls . . . saved by water," of whom St. Peter speaks. Six Pleiades only are usually seen by the naked eye; traditionally seven were seen; but the Rabbis assumed that two, not one, were lost.

Perhaps we may trace a reference to this supposed association of *Kīmah* with the Flood in the passage from Amos already quoted :—

"Seek Him that maketh the seven stars and Orion, . . . that calleth for the waters of the sea, and poureth them out upon the face of the earth: the Lord is His name."

Many ancient nations have set apart days in the late autumn in honour of the dead, no doubt because the year was then considered as dead. This season being marked by the acronical rising of the Pleiades, that group has become associated with such observances. There is, however, no reference to any custom of this kind in Scripture.

What is the meaning of the inquiry addressed to Job by the Almighty?

"Canst thou bind the sweet influences of Pleiades?"

---

[1] *Berachoth*, fol. 59, col. 1.

What was the meaning which it possessed in the thought of the writer of the book? What was the meaning which we should now put on such an inquiry, looking at the constellations from the standpoint which the researches of modern astronomy have given us?

The first meaning of the text would appear to be connected with the apparent movement of the sun amongst the stars in the course of the year. We cannot see the stars by daylight, or see directly where the sun is situated with respect to them; but, in very early times, men learnt to associate the seasons of the year with the stars which were last seen in the morning, above the place where the sun was about to rise; in the technical term once in use, with the heliacal risings of stars. When the constellations were first designed, the Pleiades rose heliacally at the beginning of April, and were the sign of the return of spring. Thus Aratus, in his constellation poem writes—

"Men mark them (*i. e.* the Pleiades) rising with the solar ray,
　The harbinger of summer's brighter day."

They heralded, therefore, the revival of nature from her winter sleep, the time of which the kingly poet sang so alluringly—

"For, lo, the winter is past,
　The rain is over and gone;
　The flowers appear on the earth;
　　The time of the singing of birds is come,
　And the voice of the turtle is heard in our land;
　　The fig-tree ripeneth her green figs,
　And the vines are in blossom,
　　They give forth their fragrance."

The constellation which thus heralded the return of this genial season was poetically taken as representing the power and influence of spring. Their "sweet influences" were those that had rolled away the gravestone of snow and ice which had lain upon the winter tomb of nature. Theirs was the power that brought the flowers up from under the turf; earth's constellations of a million varied stars to shine upwards in answer to the constellations of heaven above. Their influences filled copse and wood with the songs of happy birds. Theirs stirred anew the sap in the veins of the trees, and drew forth their re-awakened strength in bud and blossom. Theirs was the bleating of the new-born lambs; theirs the murmur of the reviving bees.

Upon this view, then, the question to Job was, in effect, "What control hast thou over the powers of nature? Canst thou hold back the sun from shining in spring-time—from quickening flower, and herb, and tree with its gracious warmth? This is God's work, year by year over a thousand lands, on a million hills, in a million valleys. What canst thou do to hinder it?"

The question was a striking one; one which must have appealed to the patriarch, evidently a keen observer and lover of nature; and it was entirely in line with the other inquiries addressed to him in the same chapter.

"Where wast thou when I laid the foundations of the earth?"

The Revised Version renders the question—

"Canst thou bind the *cluster* of the Pleiades?"

reading the Hebrew word *Ma'anaddoth*, instead of *Ma'adannoth*, following in this all the most ancient versions. On this view, Job is, in effect, asked, "Canst thou gather together the stars in the family of the Pleiades and keep them in their places?"

The expression of a chain or band is one suggested by the appearance of the group to the eye, but it is no less appropriate in the knowledge which photography and great telescopes have given us. To quote from Miss Clerke's description of the nebula discovered round the brighter stars of the Pleiades — Alcyone, Asterope, Celœno, Electra, Maia, Merope and Taygeta:—

"Besides the Maia vortex, the Paris photographs depicted a series of nebulous bars on either side of Merope, and a curious streak extending like a finger-post from Electra towards Alcyone . . . Streamers and fleecy masses of cosmical fog seem almost to fill the spaces between the stars, as clouds choke a mountain valley. The chief points of its concentration are the four stars Alcyone, Merope, Maia, and Electra; but it includes as well Celœno and Taygeta, and is traceable southward from Asterope over an arc of 1° 10'. . . . The greater part of the constellation is shown as veiled in nebulous matter of most unequal densities. In some places it lies in heavy folds and wreaths, in others it barely qualifies the darkness of the sky-ground. The details of its distribution come out with remarkable clearness, and are evidently to a large extent prescribed by the relative situations of the stars. Their lines of junction are frequently marked by nebulous rays, establishing between them, no doubt, relations of great physical importance; and masses of nebula, in numerous instances, seem as if *pulled out of shape* and drawn into festoons by the attractions of neighbouring stars. But the strangest

NEBULOSITIES OF THE PLEIADES.
Photographed by Dr. Max Wolf, Heidelberg.

exemplification of this filamentous tendency is in a fine, thread-like process, 3″ or 4″ wide, but 35′ to 40′ long, issuing in an easterly direction from the edge of the nebula about Maia, and stringing together seven stars, met in its advance, like beads on a rosary. The largest of these is apparently the occasion of a slight deviation from its otherwise rectilinear course. A second similar but shorter streak runs, likewise east and west, through the midst of the formation."[1]

Later photographs have shown that not only are the several stars of the Pleiades linked together by nebulous filaments, but the whole cluster is embedded in a nebulous net that spreads its meshes far out into space. Not only is the group thus tied or bound together by nebulous clouds, it has other tokens of forming but a single family. The movements of the several stars have been carefully measured, and for the most part the entire cluster is drifting in the same direction; a few stars do not share in the common motion, and are probably apparent members, seen in perspective projected on the group, but in reality much nearer to us. The members of the group also show a family likeness in constitution. When the spectroscope is turned upon it, the chief stars are seen to closely resemble each other; the principal lines in their spectra being those of hydrogen, and these are seen as broad and diffused bands, so that the spectrum we see resembles that of the brightest star of the heavens, Sirius.

There can be little doubt but that the leaders of the group are actually greater, brighter suns than Sirius itself. We do not know the exact distance of the Pleiades, they

[1] *The System of the Stars*, 1st edit., pp. 230-232.

are so far off that we can scarcely do more than make a guess at it; but it is probable that they are so far distant that our sun at like distance would prove much too faint to be seen at all by the naked eye. The Pleiades then would seem to be a most glorious star-system, not yet come to its full growth. From the standpoint of modern science we may interpret the "chain" or "the sweet influences" of the Pleiades as consisting in the enfolding wisps of nebulosity which still, as it were, knit together those vast young suns; or, and in all probability more truly, as that mysterious force of gravitation which holds the mighty system together, and in obedience to which the group has taken its present shape. The question, if asked us to-day, would be, in effect, "Canst thou bind together by nebulous chains scores of suns, far more glorious than thine own, and scattered over many millions of millions of miles of space; or canst thou loosen the attraction which those suns exercise upon each other, and move them hither and thither at thy will?"

CHAPTER VII

ORION

KĕsÎL, the word rendered by our translators "Orion," occurs in an astronomical sense four times in the Scriptures; twice in the Book of Job, once in the prophecy of Amos, and once, in the plural, in the prophecy of Isaiah. In the three first cases the word is used in conjunction with *Kīmah*, "the Pleiades," as shown in the preceding chapter. The fourth instance is rendered in the Authorized Version—

"For the stars of heaven and the constellations (*Kĕsīlīm*) thereof shall not give their light."

The Hebrew word *Kĕsīl* signifies "a fool," and that in the general sense of the term as used in Scripture; not merely a silly, untaught, feckless person, but a godless and an impious one. Thus, in the Book of Proverbs, Divine Wisdom is represented as appealing—

"How long, ye simple ones, will ye love simplicity? the scorners delight in their scorning, and *fools* hate knowledge?"

What constellation was known to the ancient Hebrews as "the fool"? The Seventy who rendered the Old

Testament into Greek confess themselves at fault. Once, in Amos, both *Kīmah* and *Kĕsil* are left untranslated. Instead of "Him that maketh the seven stars and Orion,"

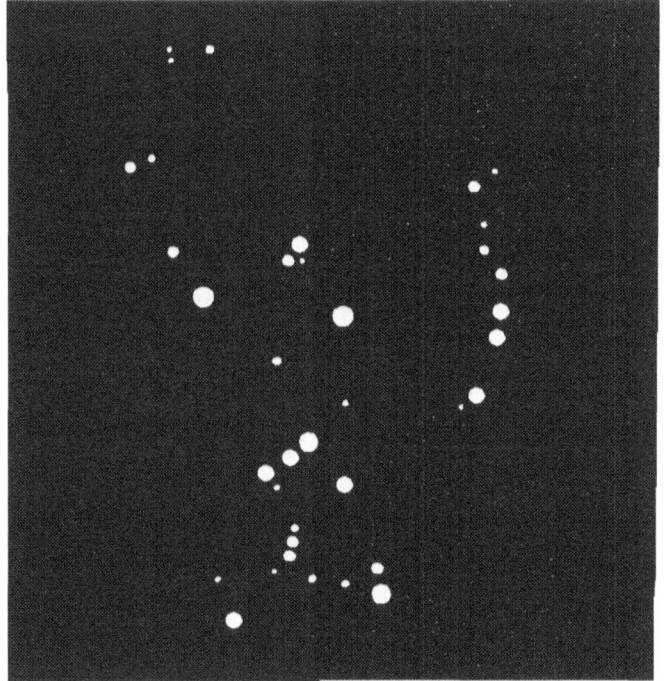

THE STARS OF ORION.

we have the paraphrase, "That maketh and transformeth all things." Once, in Job, it is rendered "Hesperus," the evening star; and in the other two instances it is given as "Orion." The tradition of the real meaning of the word as an astronomical term had been lost, or at least much confused before the Septuagint Version was undertaken. The Jews had not, so far as there is any present evidence, learned the term in Babylon, for the word has not yet been found as a star-name on any cuneiform inscription. It was well known before the Exile, for Amos and Isaiah both use it, and the fact that the author of Job also uses it, indicates that he did not gain his knowledge of the constellation during the Babylonian captivity.

The majority of translators and commentators have, however, agreed in believing that the brightest and most splendid constellation in the sky is intended—the one which we know as Orion. This constellation is one of the very few in which the natural grouping of the stars seems to suggest the figure that has been connected with it. Four bright stars, in a great trapezium, are taken to mark the two shoulders and the two legs of a gigantic warrior; a row of three bright stars, midway between the four first named, suggest his gemmed belt; another row of stars straight down from the centre star of the belt, presents his sword; a compact cluster of three stars marks his head. A gigantic warrior, armed for the battle, seems thus to be outlined in the heavens. As Longfellow describes him—

> "Begirt with many a blazing star,
> Stood the great giant, Algebar,
> Orion, hunter of the beast!
> His sword hung gleaming by his side,
> And, on his arm, the lion's hide
> Scattered across the midnight air
> The golden radiance of its hair."

In accord with the form naturally suggested by the

grouping of the stars, the Syrians have called the constellation *Gabbārā*; and the Arabs, *Al Jabbār*; and the Jews, *Gibbōr*. The brightest star of the constellation, the one in the left knee, now generally known as *Rigel*, is still occasionally called *Algebar*, a corruption of *Al Jabbār*, though one of the fainter stars near it now bears that name. The meaning in each case is "the giant," "the mighty one," "the great warrior," and no doubt from the first formation of the constellations, this, the most brilliant of all, was understood to set forth a warrior armed for the battle. There were *gibbōrim* before the Flood; we are told that after "the sons of God came in unto the daughters of men, and they bare children to them, the same became mighty men (*gibbōrim*) which were of old, men of renown."

But according to Jewish tradition, this constellation was appropriated to himself by a particular mighty man. We are told in Gen. x. that—

"Cush begat Nimrod: he began to be a mighty one (*gibbōr*) in the earth."

and it is alleged that he, or his courtiers, in order to flatter him, gave his name to this constellation, just as thousands of years later the University of Leipzic proposed to call the belt stars of Orion, *Stellæ Napoleonis*, "the Constellation of Napoleon."[1]

[1] But the fact that Napoleon's name was thus coupled with this constellation does not warrant us in asserting that Napoleon had no historical existence, and that his long contest with the great sea-power (England), with its capital on the river Thames (? *tehom*), was only a stellar myth, arising from the nearness of Orion to the Sea-monster in the sky—a variant, in fact, of the great Babylonian myth of Marduk and Tiamat, the dragon of the deep.

There was at one time surprise felt, that, deeply as the name of Nimrod had impressed itself upon Eastern tradition, his name, as such, was "nowhere found in the extensive literature which has come down to us" from Babylon. It is now considered that the word, Nimrod, is simply a Hebrew variant of Merodach, "the well-known head of the Babylonian pantheon." He was probably "the first king of Babylonia or the first really great ruler of the country." It is significant, as Mr. T. G. Pinches points out, in his *Old Testament in the Light of the Records from Assyria and Babylonia*, that just as in Genesis it is stated that "the beginning of his (Nimrod's) kingdom was Babel, and Erech, and Accad, and Calneh," so Merodach is stated, in the cuneiform records, to have built Babel and Erech and Niffer, which last is probably Calneh.

It seems necessary to make this remark, since the process of astrologizing history, whether derived from the Bible or from secular writers, has been carried very far. Thus Dr. H. Winckler writes down the account of the first three Persian kings, given us by Herodotus, as myths of Aries, Taurus, and Gemini; David and Goliath, too, are but Marduk and Tiamat, or Orion and Cetus, but David has become the Giant, and Goliath the Dragon, for "Goliath" is claimed as a word-play on the Babylonian *galittu*, "ocean." Examining an Arabic globe of date 1279 A.D.—that is to say some 4,000 years after the constellations were devised,—Dr. Winckler found that Orion was represented as left-handed. He therefore used this left-handed Orion as the link of identification between Ehud, the left-handed judge of Israel, and Tyr, the left-handed Mars of the Scandinavian pantheon. Dr. Winckler seems to have been unaware of the elementary fact that a celestial globe necessarily shows its figures "inside out." We look up to the sky, to see the actual constellations from within the sphere; we look down upon a celestial globe from without, and hence see the designs upon it as in the looking-glass.

236 THE ASTRONOMY OF THE BIBLE

The Hebrew scribes would seem to have altered the name of Merodach in two particulars: they dropped the last syllable, thus suggesting that the name was derived from *Marad*, "the rebellious one"; and they prefixed the syllable "Ni," just as "Nisroch" was written for "Assur." "From a linguistic point of view, therefore, the identifi-

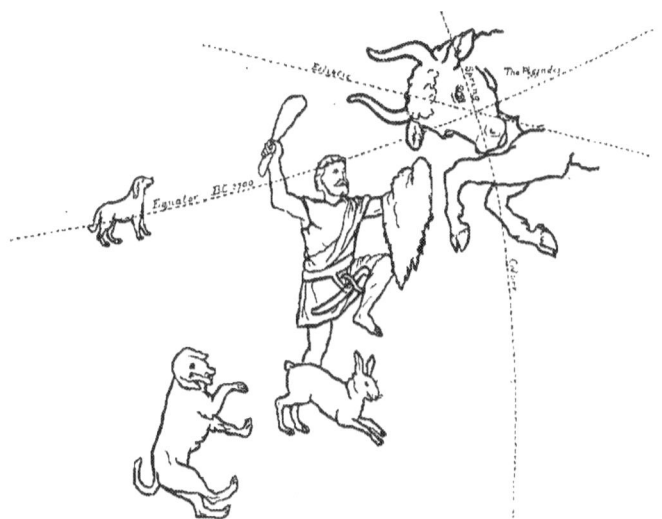

ORION AND THE NEIGHBOURING CONSTELLATIONS.

cation of Nimrod as a changed form of Merodach is fully justified."

The attitude of Orion in the sky is a striking one. The warrior is represented as holding a club in the right hand, and a skin or shield in the left. His left foot is raised high as if he were climbing a steep ascent, he seems to be endeavouring to force his way up into the zodiac, and—as Longfellow expresses it—to be beating the forehead of the

Bull. His right leg is not shown below the knee, for immediately beneath him is the little constellation of the Hare, by the early Arabs sometimes called, *Al Kursiyy al Jabbār*, " the Chair of the Giant," from its position. Behind Orion are the two Dogs, each constellation distinguished by a very brilliant star; the Greater Dog, by *Sirius*, the brightest star in the heavens; the Lesser Dog, by *Procyon*, i.e. the " Dog's Forerunner." Not far above Orion, on the shoulder of the Bull, is the little cluster of the Pleiades.

There are—as we have seen—only three passages where *Kīmah*, literally " the cluster " or " company,"—the group we know as the Pleiades,—is mentioned in Scripture; and in each case it is associated with *Kĕsīl*, " the fool,"—Orion. Several Greek poets give us the same association, likening the stars to " rock-pigeons, flying from the Hunter Orion." And Hesiod in his *Works and Days* writes—

> " Do not to plough forget,
> When the Seven Virgins, and Orion, set:
> Thus an advantage always shall appear,
> In ev'ry labour of the various year.
> If o'er your mind prevails the love of gain,
> And tempts you to the dangers of the main,
> Yet in her harbour safe the vessel keep,
> When strong Orion chases to the deep
> The Virgin stars."

There is a suggestion of intense irony in this position of Orion amongst the other constellations. He is trampling on the Hare—most timid of creatures; he is climbing up into the zodiac to chase the little company of the Pleiades—be they seven doves or seven maidens—and he is thwarted even in this unheroic attempt by the determined attitude of the guardian Bull.

A similar irony is seen in the Hebrew name for the constellation. The "mighty Hunter," the great hero whom the Babylonians had deified and made their supreme god, the Hebrews regarded as the "fool," the "impious rebel." Since Orion is Nimrod, that is Merodach, there is small wonder that *Kĕsîl* was not recognized as his name in Babylonia.[1]

The attitude of Orion—attempting to force his way upward into the zodiac—and the identification of Merodach with him, gives emphasis to Isaiah's reproach, many centuries later, against the king of Babylon, the successor of Merodach—

"Thou hast said in thine heart, I will ascend into heaven, I will exalt my throne above the stars of God: I will sit also upon the mount of the congregation, in the sides of the north: I will ascend above the heights of the clouds; I will be like the Most High."

In the sight of the Hebrew prophets and poets, Merodach, in taking to himself this group of stars, published his shame and folly. He had ascended into heaven, but his

---

[1] Dr. Cheyne says, in a note on p. 52 of *Job and Solomon*, "Heb. *K'sîl*, the name of the foolhardy giant who strove with Jehovah. The Chaldeo-Assyrian astrology gave the name *Kisiluv* to the ninth month, connecting it with the zodiacal sign Sagittarius. But there are valid reasons for attaching the Hebrew popular myth to Orion." So Col. Conder, in p. 179 of *The Hittites and their Language*, translates the name of the Assyrian ninth month, *Cisleu*, as "giant." Now Sagittarius is in the heavens just opposite to Orion, so when in the ninth month the sun was in conjunction with Sagittarius, Orion was in opposition. In *Cisleu*, therefore, the giant, Orion, was riding the heavens all night, occupying the chamber of the south at midnight, so that the ninth month might well be called the month of the giant.

glittering belt was only his fetter; he was bound and gibbeted in the sky like a captive, a rebel, and who could loose his bands?

In the thirteenth chapter of Isaiah we have the plural of *kĕsîl—kĕsîlim*. It is usually understood that we have here Orion, as the most splendid constellation in the sky, put for the constellations in general. But if we remember that *kĕsîl* stands for "Nimrod" or "Merodach," the first proud tyrant mentioned by name in Scripture, the particular significance of the allusion becomes evident—

"Behold, the day of the Lord cometh, cruel both with wrath and fierce anger, to lay the land desolate: and he shall destroy the sinners thereof out of it. For the stars of heavens and the constellations"—(that is the *kĕsîlim*, the Nimrods or Merodachs of the sky)—"thereof shall not give their light: the sun shall be darkened in his going forth, and the moon shall not cause her light to shine. And I will punish the world for their evil, and the wicked for their iniquity; and I will cause the arrogancy of the proud to cease, and will lay low the haughtiness of the terrible."

The strictly astronomical relations of Orion and the Pleiades seem to be hinted at in Amos and in Job—

"Seek Him that maketh the seven stars and Orion, and turneth the shadow of death into the morning, and maketh the day dark with night."

In this passage the parallelism seems to be between the seven stars, the Pleiades, with sunrise, and Orion with sunset. Now at the time and place when the constellations were mapped out, the Pleiades were the immediate heralds of sunrise, shortly after the spring equinox, at the season which would correspond to the early part of April in our present calendar. The rising of Orion at sunset—his acronical rising—was early in December, about the time when the coldest season of the year begins. The astronomical meaning of the "bands of Orion" would therefore be the rigour in which the earth is held during the cold of winter.

It is possible that the two great stars which follow Orion, *Sirius* and *Procyon*, known to the ancients generally and to us to-day as "the Dogs," were by the Babylonians known as "the Bow-star" and "the Lance-star"; the weapons, that is to say, of Orion or Merodach. Jensen identifies Sirius with the Bow-star, but considers that the Lance-star was Antares; Hommel, however, identifies the Lance-star with Procyon. In the fifth tablet of the Babylonian Creation epic as translated by Dr. L. W. King, there is an interesting account of the placing of the Bow-star in the heavens. After Merodach had killed Tiamat—

75. "The gods (his fathers) beheld the net which he had made,
76. They beheld the bow and how (its work) was accomplished.
77. They praised the work which he had done [. . .]
78. Then Anu raised [the . . .] in the assembly of the gods.
79. He kissed the bow, (saying), 'It is [. . .]'!
80. And thus he named the names of the bow, (saying),
81. '*Long-wood* shall be one name, and the second name [shall be . . .],
82. And its third name shall be the *Bow-star*, in heaven [shall it . . .]!'
83. Then he fixed a station for it."

Dr. Cheyne even considers that he has found a reference to these two stars in Job xxxviii. 36—

"Who hath put wisdom in the inward parts (Lance-star),
 Or who hath given understanding to the heart (Bow-star)."

But this interpretation does not appear to have been generally accepted. The same high authority suggests that the astronomical allusions in Amos may have been inserted by a post-exilic editor, thus accounting for the occurrence of the same astronomical terms as are found in Job, which he assigns to the exilic or post-exilic period. This seems a dangerous expedient, as it might with equal reason be used in many other directions. Further, it entirely fails to explain the real difficulty that *kīmah* and *kĕsīl* have not been found as Babylonian constellation names, and that their astronomical signification had been lost by the time that the "Seventy" undertook their labours.

Quite apart from the fact that the Babylonians could not give the name of "Fool" to the representation in the sky of their supreme deity, the Hebrews and the Babylonians regarded the constellation in different ways. Several Assyriologists consider that the constellations, *Orion* and *Cetus*, represent the struggle between Merodach and Tiamat, and this conjecture is probably correct, so far as Babylonian ideas of the constellations are concerned, for Tiamat is expressly identified on a Babylonian tablet with a constellation near the ecliptic.[1] But this means that the myth originated in the star figures, and was the Babylonian interpretation of them. In this case, Cetus—that is Tiamat—must have been considered as a goddess, and as directly and immediately the ancestress of

[1] Dr. L. W. King, *Tablets of Creation*, appendix iii. p. 208.

all the gods. Orion—Merodach—must have been likewise a god, the great-great-grandson of Tiamat, whom he destroys.

The Hebrew conception was altogether different. Neither Merodach, nor Tiamat, nor the constellations of Orion and Cetus, nor the actual stars of which they are composed, are anything but creatures. Jehovah has made Orion, as well as the "Seven Stars," as "His hand hath formed the crooked serpent." By the mouth of Isaiah He says, "I form the light, and create darkness: I make peace, and create evil: I the Lord, do all these things." The Babylonian view was of two divinities pitted against each other, and the evil divinity was the original and the originator of the good. In the Hebrew view, even the powers of evil are created things; they are not self-existent.

And the Hebrews took a different view from the Babylonians of the story told by these constellations. The Hebrews always coupled Orion with the Pleiades; the Babylonians coupled Orion with Cetus—that is, Merodach with Tiamat.

The view that has come down to us through the Greeks agrees much better with the association of the constellations as held amongst the Hebrews, rather than amongst the Babylonians. The Hunter Orion, according to the Greeks, chased the Pleiades—the little company of Seven Virgins, or Seven Doves—and he was confronted by the Bull. In their view, too, the Sea-monster was not warring against Orion, but against the chained woman, Andromeda.

CHAPTER VIII

MAZZAROTH

WE have no assistance from any cuneiform inscriptions as to the astronomical significance of *'Ayish, Kīmah,* and *Kĕsīl,* but the case is different when we come to *Mazzaroth.* In the fifth tablet of the Babylonian Creation epic we read—

"1. He (Marduk) made the stations for the great gods;
2. The stars, their images, as the stars of the zodiac, he fixed.
3. He ordained the year, and into sections (*mizrāta*) he divided it;
4. For the twelve months he fixed three stars.
5. After he had [. . .] the days of the year [. . .] images
6. He founded the station of Nibir to determine their bounds;
7. That none might err or go astray.
8. He set the station of Bēl and Ea along with him."

In the third line *mizrāta,* cognate with the Hebrew *Mazzārōth,* means the sections or divisions of the year, corresponding to the signs of the zodiac mentioned in the second line. There can therefore be little doubt that the translators who gave us our English versions are practically correct in the rendering of Job xxxviii. 32 which they give in the margin, "Canst thou bring forth Mazzaroth (*or* the twelve signs) in his season?"

The foregoing extract from the fifth tablet of Creation

has no small astronomical interest. Merodach is represented as setting in order the heavenly bodies. First of all he allots their stations to the great gods, dividing to them the constellations of the zodiac, and the months of the year; so that the arrangement by which every month had its tutelary deity or deities, is here said to be his work. Next, he divides up the constellations of the zodiac; not merely arranging the actual stars, but appropriating to each constellation its special design or "image." Third, he divides up the year to correspond with the zodiac, making twelve months with three "stars" or constellations to each. In other words, he carries the division of the zodiac a step further, and divides each sign into three equal parts, the "decans" of the astrologers, each containing 10° (*deka*) of the ecliptic.

The statement made in line 4 refers to an important development of astronomy. The *constellations* of the zodiac, that is, the groups made up of the actual stars, are very unequal in size and irregular in shape. The numerous theories, ancient or modern, in which the constellations are supposed to owe their origin to the distinctive weather of the successive months, each constellation figure being a sort of hieroglyph for its particular month, are therefore all manifestly erroneous, for there never could have been any real fixed or steady correlation between the constellations and the months. Similarly, the theories which claim that the ancient names for the months were derived from the constellations are equally untenable. Some writers have even held both classes of theory, overlooking the fact that they mutually contradict each other.

But there came a time when the inconvenience of the unequal division of the zodiac by the constellations was felt to be an evil, and it was remedied by dividing the ecliptic into twelve equal parts, each part being called after the constellation with which it corresponded most nearly at the time such division was made. These equal divisions have been called the *Signs* of the zodiac. It must be clearly understood that they have always and at all times been imaginary divisions of the heavens, that they were never associated with real stars. They were simply a picturesque mode of expressing celestial longitude; the distance of a star from the place of the sun at the spring equinox, as measured along the ecliptic,—the sun's apparent path during the year.

The Signs once arranged, the next step was an easy one. Each sign was equivalent to 30 degrees of longitude. A third of a sign, a "decan," was 10 degrees of longitude, corresponding to the "week" of ten days used in Egypt and in Greece.

This change from the constellations to the Signs cannot have taken place very early. The place of the spring equinox travels backwards amongst the stars at the rate of very little more than a degree in 72 years. When the change was made the spring equinox was somewhere in the constellation *Aries*, the Ram, and therefore Aries was then adopted as the first Sign, and must always remain such, since the Signs move amongst the stars with the equinox.

We cannot fix when this change was made within a few years, but it cannot have been *before* the time when the sun at the spring equinox was situated just below *Hamal*, the brightest star of the Ram. This was about 700 B.C. The equal division of the zodiac must

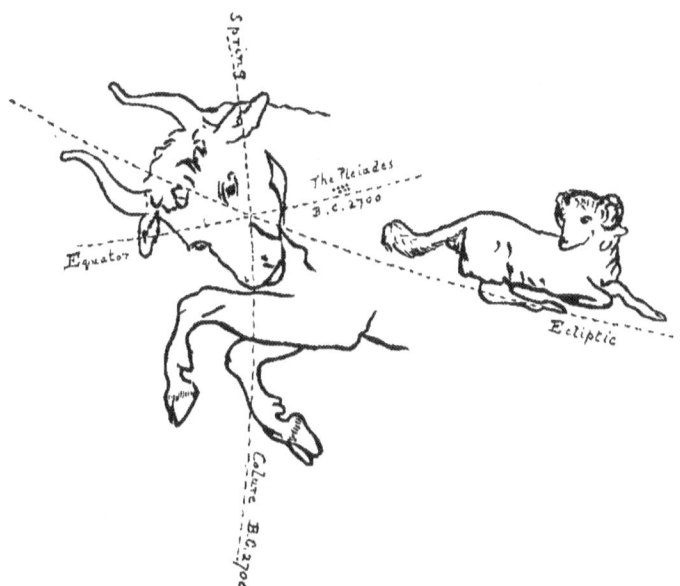

POSITION OF SPRING EQUINOX, B.C. 2700.

have taken place not earlier than this, and with it, the Bull must have been deposed from the position it had always held up to that time, of leader of the zodiac. It is probable that some direct method of determining the equinox itself was introduced much about the same time. This new system involved nothing short of a revolution in astronomy, but the Babylonian Creation story

MAZZAROTH

implies that this revolution had already taken place when it was composed, and that the equal division of the zodiac was already in force. It is possible that the sixth and seventh lines of the poem indicate that the Babylonians had already noticed a peculiar fact, viz. that just as the moon passes through all the signs in a month, whilst the sun passes through only one sign in that time; so the sun passes through all the signs in a year, whilst

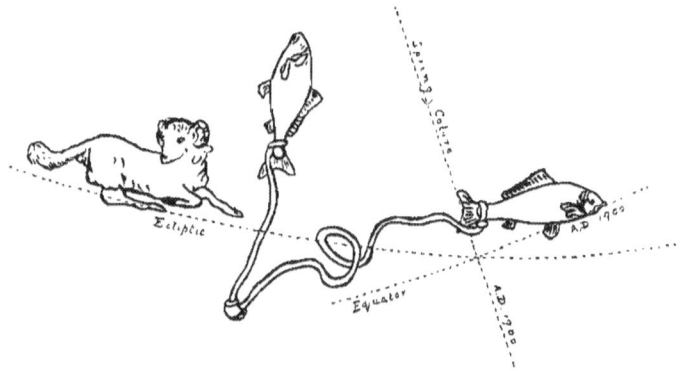

POSITION OF SPRING EQUINOX, A.D. 1900.

Jupiter passes through but one sign. *Nibir* was the special Babylonian name of the planet Jupiter when on the meridian; and Merodach, as the deity of that planet, is thus represented as pacing out the bounds of the zodiacal Signs by his movement in the course of the year. The planet also marks out the third part of a sign, *i.e.* ten degrees; for during one-third of each year it appears to retrograde, moving from east to west amongst the stars instead of from west to east. During this retrogression it covers the breadth of one "decan" = ten degrees.

The Babylonian Creation epic is therefore quite late, for it introduces astronomical ideas not current earlier than 700 B.C. in Babylonia or anywhere else. This new development of astronomy enables us also to roughly date the origin of the different orders of systematic astrology.

Astrology, like astronomy, has passed through successive stages. It began at zero. An unexpected event in the heavens was accounted portentous, because it was unexpected, and it was interpreted in a good or bad sense according to the state of mind of the beholder. There can have been at first no system, no order, no linking up of one specific kind of prediction with one kind of astronomical event. It can have been originally nothing but a crude jumble of omens, just on a level with the superstitions of some of our peasantry as to seeing hares, or cats, or magpies; and the earliest astrological tablets from Mesopotamia are precisely of this character.

But the official fortune-tellers at the courts of the kings of Nineveh or Babylon must speedily have learned the necessity of arranging some systems of prediction for their own protection—systems definite enough to give the astrologer a groundwork for a prediction which he could claim was dependent simply upon the heavenly bodies, and hence for which the astrologer could not be held personally responsible, and at the same time elastic enough to enable him to shape his prediction to fit in with his patron's wishes. The astrology of to-day shows the same essential features.

This necessity explains the early Babylonian tablets with catalogues of eclipses on all days of the month, and in all quarters of the sky. The great majority of the eclipses could never happen, but they could be, none-the-less, made use of by a court magician. If an eclipse of the sun took place on the 29th day and in the south, he could always point out how exceedingly unpleasant things might have been for the king and the country if he, the magician, had not by his diligence, prevented its happening, say, on the 20th, and in the north. A Zulu witch-doctor is quite equal to analogous subterfuges to-day, and no doubt his Babylonian congeners were not less ingenious 3,000 years ago. Such subterfuges were not always successful when a Chaka or a Nebuchadnezzar had to be dealt with, but with kings of a more ordinary type either in Zululand or Mesopotamia they would answer well enough.

Coming down to times when astronomy had so far advanced that a catalogue of the stars had been drawn up, with their positions determined by actual measurement, it became possible for astrologers to draw up something like a definite system of prediction, based upon the constellations or parts of a constellation that happened to be rising at any given moment, and this was the system employed when Zeuchros of Babylon wrote in the first century of our era. His system must have been started later than 700 B.C., for in it Aries is considered as the leader of the zodiac; the constellations are already disestablished in favour of the Signs; and the Signs are each divided into three. A practical drawback to this particular astrological system was that the aspect presented by the heavens on one evening was precisely the same as that presented on the next evening four minutes earlier. The field for prediction therefore was very limited and repeated itself too much for the purpose of fortune-tellers.

The introduction of the planets into astrology gave a greater diversity to the material used by the fortune-tellers. An early phase of planetary astrology consisted in the allotment of a planet to each hour of the day and also to each day of the week. It has been already shown in the chapter on " Saturn and Astrology," that this system arose from the Ptolemaic idea of the solar system grafted on the Egyptian division of the day into twenty-four hours, and applied to the week of seven days. It probably originated in Alexandria, and arose not earlier than the third century before our era. Mathematical astrology—the complex system now in vogue—involves a considerable knowledge of the apparent movements of the planets and a development of mathematics such as did not exist until the days of Hipparchus. It also employs the purely imaginary signs of the zodiac, not the constellations; and reckons the first point of Aries as at the spring equinox. So far as we can ascertain, the spring equinox marked the first point of the constellation Aries about B.C. 110.

All these varied forms of astrology are therefore comparatively recent. Before that it was of course reckoned ominous if an eclipse took place, or a comet was seen, or a bright planet came near the moon, just as spilling salt or crossing knives may be reckoned ominous to-day. The omens had as little to do with observation, or with anything that could be called scientific, in the one case as in the other.

It is important to realize that astrology, as anything more than the crude observance of omens, is younger than astronomy by at least 2,000 years.

Mazzārōth occurs only once in the Bible, viz. in Job xxxviii. 32, already so often quoted, but a similar word *Mazzālōth* occurs in 2 Kings xxiii. 5, where it is said that Josiah put down the idolatrous priests, "them also that burned incense unto Baal, to the sun, and to the moon, and to the planets (*Mazzālōth*), and to all the host of heaven." The context itself, as well as the parallel passage in Deuteronomy—" When thou seest the sun, and the moon, and the stars, even all the host of heaven, shouldst be driven to worship them,"—shows clearly that celestial luminaries of some kind are intended, probably certain groups of stars, distinguished from the general "host of heaven."

Comparing Job ix. 9, with Job xxxviii. 31, 32, we find '*Ash*, or '*Ayish*, *Kīmah* and *Kĕsil* common to the two passages; if we take '*Ash* and '*Ayish* as identical, this leaves the "chambers of the south" as the equivalent of *Mazzaroth*. The same expression occurs in the singular in Job xxxvii. 9—" Out of the south (*marg.* chamber) cometh the whirlwind." There need be but little question as to the significance of these various passages. The correspondence of the word *Mazzārōth* with the Babylonian *mizrātā*, the "divisions" of the year, answering to the twelve signs of the zodiac, points in exactly the same direction as the correspondence in idea which is evident between the "chambers of the south" and the Arabic *Al manāzil*, "the mansions" or "resting-places" of the moon in the lunar zodiac.

Mazzaroth are therefore the "divisions" of the zodiac, the "chambers" through which the sun successively passes in the course of the year, his "resting-place" for a month. They are "the chambers of the south," since that is their distinctive position. In Palestine, the sun, even at rising or setting at midsummer, passes but little to the north of east or west. Roughly speaking, the "south" is the sun's quarter, and therefore it is necessarily the quarter of the constellation in which the sun is placed.

It has been made an objection to this identification that the Israelites are said to have worshipped *Mazzālōth*, and we have no direct evidence that the signs or constellations of the zodiac were worshipped as such. But this is to make a distinction that is hardly warranted. The Creation tablets, as we have seen, distinctly record the allocation of the great gods to the various signs, Merodach himself being one of the three deities associated with the month Adar, just as in Egypt a god presided over each one of the thirty-six decades of the year.

Again, it is probable that the "golden calf," worshipped by the Israelites in the wilderness, and, after the disruption, at Bethel and at Dan, was none other than an attempt to worship Jehovah under the symbol of Taurus, the leader of the zodiac and cognizance of the tribe of Joseph; regarded as a type of Him Who had been the Leader of the people out of Egypt, and the Giver of the blessings associated with the return of the sun to Taurus, the revival of nature in spring-time. It was intended as a worship of Jehovah; it was in reality dire rebellion against Him, and a beginning of the worship of "*Mazzālōth* and the heavenly host;" an idolatry that was bound to bring other idolatries in its train.

A three-fold symbol found continually on Babylonian monuments, "the triad of stars," undoubtedly at one time set forth Sin, the moon-god, Samas, the sun-god, and Ištar, in this connection possibly the planet Venus. It has therefore been suggested by Prof. Schiaparelli that *Mazzālōth* is the planet Venus; and, since the word is plural in form, Venus in her double capacity;—sometimes an evening, sometimes a morning star. The sun and the moon and *Mazzālōth* would then set forth the three brightest luminaries, whilst the general congress of stars would be represented by the "host of heaven." But though Venus is sometimes the brightest of the planets, she is essentially of the same order as Jupiter or Mars, and is not of the same order as the sun and moon, with whom, on this supposition, she is singled out to be ranked. Moreover, if Ištar or Ashtoreth were intended in this passage, it does not appear why she should not be expressly named as such; especially as Baal, so often coupled with her, is named. The "triad of stars," too, had originally quite a different meaning, as will be seen later.

Moreover, the parallelism between Job ix. and Job xxxviii. is destroyed by this rendering, since the planet Venus could not be described as "the chambers of the south." These are therefore referred by Professor Schiaparelli to the glorious mass of stars in the far south, shining in the constellations that set forth the Deluge story,—the Ship, and the Centaur, much the most brilliant region of the whole sky.

Another interpretation of *Mazzaroth* is given by Dr. Cheyne, on grounds that refute Professor Schiaparelli's suggestion, but it is itself open to objection from an astronomical point of view. He writes—

"*Mazzaroth* is probably not to be identified with *Mazzaloth* (2 Kings xxiii. 5) in spite of the authority of the Sept. and the Targum. . . . *Mazzaroth* = Ass. *Mazarati*; *Mazzaloth* (i.e. the zodiacal signs) seems to be the plural of *Mazzāla* = Ass. *Manzaltu*, station." [1]

Dr. Cheyne therefore renders the passage thus—

"Dost thou bring forth the moon's watches at their season,
And the Bear and her offspring—dost thou guide them?
Knowest thou the laws of heaven?
Dost thou determine its influence upon the earth?"

Mazzaloth are therefore "the zodiacal signs," but *Mazzaroth* "the watches or stations of the moon, which marked the progress of the month;" [2] or, in other words, the lunar zodiac.

But the lunar and the solar zodiac are only different ways of dividing the same belt of stars. Consequently when, as in the passage before us, reference is made to the actual belt of stars as a whole, there is no difference between the two. So that we are obliged, as before, to consider *Mazzaroth* and *Mazzaloth* as identical, and both as setting forth the stars of the zodiac.

[1] Rev. T. K. Cheyne, M.A., *Job and Solomon*, p. 290.
[2] *Ibid.*, p. 52.

So far as the two zodiacs differ, it is the solar and not the lunar zodiac that is intended. This is evident when we consider the different natures of the apparent motions of the sun and the moon. The sun passes through a twelfth part of the zodiac each month, and month by month the successive constellations of the zodiac are brought out, each in its own season; each having a period during which it rises at sunset, is visible the whole night, and sets at sunrise. The solar *Mazzaroth* are therefore emphatically brought out, each "in its season." Not so the lunar *Mazzaroth*.

The expression, "the watches or stations of the moon which marked the progress of the month," is unsuitable when astronomically considered. "Watches" refer strictly to divisions of the day and night; the "stations" of the moon refer to the twenty-seven or twenty-eight divisions of the lunar zodiac; the "progress of the month" refers to the complete sequence of the lunar phases. These are three entirely different matters, and Dr. Cheyne has confused them. The progress of the moon through its complete series of stations is accomplished in a siderial month—that is, twenty-seven days eight hours, but from the nature of the case it cannot be said that these "stations" are brought out each in his season, in that time, as a month makes but a small change in the aspect of the sky. The moon passes through the complete succession of its phases in the course of a synodical month, which is in the mean twenty-nine days, thirteen hours—that is to say from new to new, or full to full—but no particular star, or constellation, or "station" has any fixed relation to any one given phase of the moon. In the course of some four or five years the moon will have been both new and full in every one of the "lunar stations."

"Knowest thou the ordinances of heaven?
Canst thou set the dominion thereof in the earth?"

He, who has lived out under the stars, in contact with the actual workings of nature, knows what it is to watch "Mazzaroth" brought "out in his season;" the silent return to the skies of the constellations, month by month, simultaneous with the changes on the face of the earth. Overhead, the glorious procession, so regular and unfaltering, of the silent, unapproachable stars: below, in unfailing answer, the succession of spring and summer, autumn and winter, seedtime and harvest, cold and heat, rain and drought. If there be but eyes to see, this majestic Order, so smooth in working, so magnificent in scale, will impress the most stolid as the immediate acting of God; and the beholder will feel at the same a reverent awe, and an uplifting of the spirit as he sees the action of "the ordinances of heaven," and the evidence of "the dominion thereof in the earth."

Dr. Cheyne, however, only sees in these beautiful and appropriate lines the influence upon the sacred writer of "the physical theology of Babylonia";[1] in other words, its idolatrous astrology, "the influence of the sky upon the earth."

But what would Job understand by the question, "Canst thou bring forth Mazzārōth in his season?" Just

[1] Rev. T. K. Cheyne, M.A., *Job and Solomon*, p. 52.

MAZZAROTH

this: "Canst thou so move the great celestial sphere that the varied constellations of the zodiac shall come into view, each in their turn, and with them the earth pass through its proper successive seasons?" The question therefore embraced and was an extension of the two that preceded it. "Canst thou bind the sweet influences of the Pleiades? Canst thou prevent the revival of all the forces of nature in the springtime?" and "Canst thou loose the bands of Orion; canst thou free the ground from the numbing frosts of winter?"

The question to us would not greatly differ in its meaning, except that we should better understand the mechanism underlying the phenomena. The question would mean, "Canst thou move this vast globe of the earth, weighing six thousand million times a million million tons, continually in its orbit, more than 580 millions of miles in circuit, with a speed of nearly nineteen miles in every second of time, thus bringing into view different constellations at different times of the year, and presenting the various zones of the earth in different aspects to the sun's light and heat?" To us, as to Job, the question would come as:

"Knowest thou the ordinances of heaven?
Canst thou set the dominion thereof in the earth?"

It is going beyond astronomy, yet it may be permitted to an astronomer, to refer for comparison to a parallel thought, not couched in the form of a question, but in the form of a prayer:

"Thy will be done,
As in heaven, so in earth."

CHAPTER IX

ARCTURUS

IN two passages of the Book of Job a word, *'Ash* or *'Ayish*, is used, by context evidently one of the constellations of the sky, but the identification of which is doubtful. In our Authorized Version the first passage is rendered thus:—

(God) "Which maketh Arcturus (*'Ash*), Orion, and Pleiades, and the chambers of the south";

and the second:—

"Canst thou bind the sweet influences of Pleiades,
Or loose the bands of Orion?
Canst thou bring forth Mazzaroth in his season?
Or canst thou guide Arcturus (*'Ayish*) with his sons?
Knowest thou the ordinances of heaven?
Canst thou set the dominion thereof in the earth?"

The words (or word, for possibly *'Ayish* is no more than a variant of *'Ash*) here translated "Arcturus" were rendered by the "Seventy" as "Arktouros" in the first passage; as "Hesperos" in the second passage; and their rendering was followed by the Vulgate. The rendering Hesper or Vesper is absurd, as "the sons" of Hesper has no meaning. "Arktouros" is not improbably a misrendering of "Arktos," "the north," which would give a free but not a literal translation of the meaning of the passage. In another passage from Job (xxxvii. 9) where the south wind is contrasted with the cold from another quarter of the sky, the "Seventy"—again followed by the Vulgate—rendered it as "cold from Arcturus." Now cold came to the Jews, as it does to us, from the north, and the star which we know as Arcturus could not be described as typifying that direction either now or when the Septuagint or Vulgate versions were made. The Peschitta, the Syriac version of the Bible, made about the second century after Christ, gives as the Syriac equivalent for ʻAsh, or ʻAyish, the word *ʻiyūthā*, but it also renders *Kĕsîl* by the same word in Amos v. 8, so that the translators were evidently quite at sea as to the identity of these constellations. We are also in doubt as to what star or constellation the Syrians meant by *ʻIyūthā*, and apparently they were in some doubt themselves, for in the Talmud we are told that there was a disputation, held in the presence of the great teacher Rabbi Jehuda, about 150 years after Christ, whether *ʻIyūthā* was situated in the head of the Bull, or in the tail of the Ram. Oriental scholars now assign it either to Aldebaran in the head of the Bull, the "sons" being in this case the other members of the Hyades group of which Aldebaran is the brightest star; or else identifying it with the Arabic *el-ʻaiyūq*, the name of the star which the Greeks call *Aix*, and we call Capella, the "sons" on this inference being the three small stars near, called by the Greeks and by ourselves the "Kids." The word '*Ash* is used several times in Scripture, but without any astronomical signification, and is there rendered "moth," as in Isaiah, where it says—

"Lo, they all shall wax old as a garment; the moth ('*Ash*) shall eat them up."

This literal significance of the word does not help, as we know of no constellation figured as a "moth" or bearing any resemblance to one.

But the word '*ash*, or '*ayish* does not differ importantly from the word *na'sh*, in Hebrew "assembly," in Arabic "bier," which has been the word used by the Arabs from remote antiquity to denote the four bright stars in the hind-quarters of the Great Bear; those which form the body of the Plough. Moreover, the three stars which form the "tail"of the Great Bear," or the "handle" of the Plough have been called by the Arabs *benāt na'sh*, "the daughters of na'sh." The Bear is the great northern constellation, which swings constantly round the pole, always visible throughout the changing seasons of the year. There should be no hesitation then in accepting the opinion of the Rabbi, Aben Ezra, who saw in '*Ash*, or '*Ayish* the quadrilateral of the great Bear, whose four points are marked by the bright stars, Alpha, Beta, Gamma and Delta, and in the "sons" of '*Ayish*, the three stars, Epsilon, Zeta, and Eta. Our Revised Version therefore renders the word as "Bear."

In both passages of Job, then, we get the four quarters of the sky marked out as being under the dominion of the Lord. In the ninth chapter they are given in the order—

The Bear, which is in the North;
Orion, in its acronical rising, with the sun setting in the West;
The Pleiades, in their heliacal rising, with the sun rising in the East;
And the Chambers of the South.

In the later passage they are given with fuller illustration, and in the order—

The Pleiades, whose "sweet influences" are given by their heliacal rising in spring time, with the sun rising in the East;
Orion, whose "bands" are those of winter, heralded by his acronical rising with the sun setting in the West;
Mazzaroth, the constellations of the zodiac corresponding to the Chambers of the South, which the sun occupies each in its "season."
The Bear with its "sons," who, always visible, are unceasingly guided round the pole in the North.

The parallelism in the two passages in Job gives us the right to argue that '*Ash* and '*Ayish* refer to the same constellation, and are variants of the same name; possibly their vocalization was the same, and they are but two divergent ways of writing the word. We must therefore reject Prof. Schiaparelli's suggestion made on the authority of the Peschitta version of the Scriptures and of Rabbi Jehuda, who lived in the second century A.D., that '*Ash* is '*Iyūthā* which is Aldebaran, but that '*Ayish* and his "sons" may be Capella and her "Kids."

Equally we must reject Prof. Stern's argument that *Kīmah* is Sirius, *Kĕsīl* is Orion, *Mazzārōth* is the Hyades and '*Ayish* is the Pleiades. He bases his argument on the order in which these names are given in the second passage of Job, and on the contention of Otfried Müller that there are only four out of the remarkable groups of stars placed in the middle and southern regions of the sky which have given rise to important legends in the primitive mythology of the Greeks. These groups follow one after the other in a belt in the sky in the order just given, and their risings and settings were important factors in the old Greek meteorological and agricultural calendars. Prof. Stern assumes that *kĕsîl* means Orion, and from this identification deduces the others, neglecting all etymological or traditional evidences to the contrary. He takes no notice of the employment of the same names in passages of Scripture other than that in the thirty-eighth chapter of Job. Here he would interpret the "chain," or "sweet influences" of *Kîmah* = "Sirius the dog," by assuming that the Jews considered that the dog was mad, and hence was kept chained up. More important still, he fails to recognize that the Jews had a continental climate in a different latitude from the insular climate of Greece, and that both their agricultural and their weather conditions were different, and would be associated with different astronomical indications.

In the 9th verse of the 37th chapter of Job we get an antithesis which has already been referred to—

"Out of the south cometh the whirlwind : and cold out of the north."

The Hebrew word here translated "north" is *mezarîm*, a plural word which is taken literally to mean "the scatterings." For its interpretation Prof. Schiaparelli

ARCTURUS

makes a very plausible suggestion. He says, "We may first observe that the five Hebrew letters with which this name was written in the original unpointed text could equally well be read, with a somewhat different pointing, as *mizrim*, or also as *mizrayim*, of which the one is the plural, the other the dual, of *mizreh*. Now *mizreh* means a winnowing-fan, the instrument with

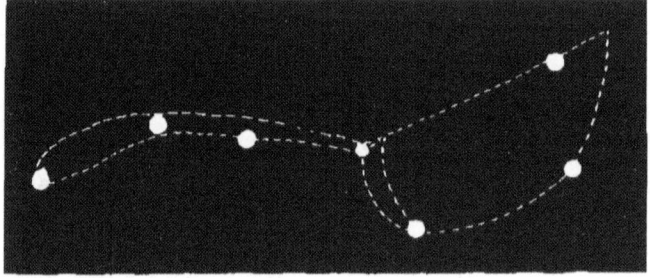

STARS OF THE PLOUGH, AS THE WINNOWING FAN.

which grain is scattered in the air to sift it; and it has its root, like *mezarim*, in the word *zarah*, . . . which, besides the sense *dispersit*, bears also the sense *expandit, ventilavit*."[1]

If Prof. Schiaparelli is correct in his supposition, then the word translated "north" in our versions is literally the "two winnowing fans," names which from the form suggested by the stars we may suppose that the Jews gave to the two Bears in the sky, just as the Chinese called them the "Ladles," and the Americans call them the "Big Dipper" and the "Little Dipper." The sense is still that of the north, but we may recognize in the

[1] *Astronomy in the Old Testament*, p. 69.

word employed another Jewish name of the constellation, alternative with 'Ash or 'Ayish, or perhaps used in order to include in the region the Lesser as well as the Greater Bear. We should not be surprised at finding an alternative name for this great northern constellation, for we ourselves call it by several different appellations, using them indiscriminately, perhaps even in the course of a single paragraph.

What to Job did the question mean which the Lord addressed to him: "Canst thou guide the Bear and his sons?" To Job it meant, "Canst thou guide this great constellation of stars in the north, in their unceasing round, as a charioteer guides his horses in a wide circle, each keeping to his proper ring, none entangling himself with another, nor falling out of his place?"

What would the same question mean to us, if addressed to us to-day? In the first place we might put it shortly as "Canst thou turn the earth on its axis regularly and continuously, so as to produce this motion of the stars round the pole, and to make day and night?" But modern astronomy can ask the question in a deeper and a wider sense.

It was an ancient idea that the stars were fixed in a crystal sphere, and that they could not alter their relative positions; and indeed until the last century or two, instruments were not delicate enough to measure the small relative shift that stars make. It is within the last seventy years that we have been able to measure the "annual parallax" of certain stars,—that is, the difference in the position of a star when viewed by the earth from the opposite ends of a diameter of the earth's orbit round the sun. Besides their yearly shift due to "annual parallax," most stars have a "proper," or "peculiar motion" of their own, which is in most cases a very small amount indeed, but can be determined more easily than "annual parallax" because its effect accumulates year after year. If, therefore, we are able to observe a star over a period of fifty, or a hundred or more years, it may seem to have moved quite an appreciable amount when examined by the powerful and delicate instruments that we have now at our disposal. Observations of the exact positions of stars have been made ever since the founding of Greenwich Observatory, so that now we have catalogues giving the "proper motions" of several hundreds of stars. When these are examined it is seen that some groups of stars move in fellowship together through space, having the same direction, and moving at the same rate, and of these companies the most striking are the stars of the Plough, that is '*Ayish* and his sons. Not all the stars move together; out of the seven, the first and the last have a different direction, but the other five show a striking similarity in their paths. And not only are their directions of movement, and the amounts of it, the same for the five stars, but spectroscopic observations of their motion in the line of sight show that they are all approaching us with a speed of about eighteen miles a second, that is to say with much the same speed as the earth moves in her orbit round the sun. Another indication of their "family likeness" is that all their spectra are similar. A German astronomer, Dr. Höffler, has found for this system a distance from us so great that it would take light 192 years to travel from them to us. Yet so vast is this company of five stars that it would take light seventy years, travelling at the rate of 186,000 miles in every second of time to go from the leading star, *Merak*—Beta of the Bear—to *Mizar*—Zeta of the Bear—the final brilliant of the five. So bright and great are these suns that they shine to us as gems of the second magnitude, and yet if our sun were placed amongst them at their distance from us he would be invisible to the keenest sight.

Dr. Höffler's estimate may be an exaggerated one, but it still remains true that whilst the cluster of the Pleiades forms a great and wonderful family group, it is dwarfed into insignificance by the vast distances between these five stars of the Great Bear. Yet these also form one family, though they are united by no nebulous bands, and are at distances so great from each other that the bonds of gravitation must cease to show their influence; yet all are alike, all are marshalled together in their march under some mysterious law. We cannot answer the question, "By what means are '*Ayish* and his sons guided?" much more are we speechless when we are asked, "Canst thou guide them?"

"BLOW UP THE TRUMPET IN THE NEW MOON."

BOOK III

TIMES AND SEASONS

CHAPTER I

THE DAY AND ITS DIVISIONS

THERE is a difference of opinion at the present day amongst astronomers as to the time in which the planet Venus rotates upon her axis. This difference arises through the difficulty of perceiving or identifying any markings on her brilliantly lighted surface. She is probably continually cloud-covered, and the movements of the very faint shadings that are sometimes seen upon her have been differently interpreted. The older observers concurred in giving her a rotation period of $23^h\ 21^m$, which is not very different from that of the earth. Many astronomers, amongst them Schiaparelli, assign a rotation period of 225 days, that is to say the same period as that in which she goes round the sun in her orbit. The axis on which she rotates is almost certainly at right angles to the plane in which she moves round the sun, and she has no moon.

We do not know if the planet is inhabited by intelligent beings, but assuming the existence of such, it will be instructive to inquire as to the conditions under which they must live if this view be correct, and the rotation period of Venus, and her revolution period be the same.

Venus would then always turn the same face to the sun, just as our moon always turns the same face to us and so never appears to turn round. Venus would therefore have no "days," for on her one hemisphere there would be eternal light, and on the other eternal darkness. Since she has no moon, she has no "month." Since she moves round the sun in a circle, and the axis through her north and south poles lies at right angles to her ecliptic, she has no "seasons," she can have no "year." On her daylight side, the sun remains fixed in one spot in the sky, so long as the observer does not leave his locality; it hangs overhead, or near some horizon, north, south, east, or west, continually. There are no "hours," therefore no divisions of time, it might be almost said no "time" itself. There are no points of the compass even, no north, south, east or west, no directions except towards the place where the sun is overhead or away from it. There could be no history in the sense we know it, for there would be no natural means of dating. "Time" must there be artificial, uncertain and arbitrary.

On the night side of Venus, if her men can see the stars at all for cloud, they would perceive the slow procession of stars coming out, for Venus turns continually to the heavens—though not to the sun. *Mazzaroth* would still be brought out in his season, but there would be no answering change on Venus. Her men might still know the ordinances of heaven, but they could not know the dominion thereof set upon their earth.

THE DAY AND ITS DIVISIONS

This imaginary picture of the state of our sister planet may illustrate the fourteenth verse of the first chapter of Genesis:—

"And God said, Let there be lights in the firmament of the heaven to divide the day from the night; and let them be for signs, and for seasons, and for days, and years."

The making of the calendar is in all nations an astronomical problem: it is the movements of the various heavenly bodies that give to us our most natural divisions of time. We are told in Deuteronomy:—

"The sun, and the moon, and the stars, even all the host of heaven, . . . the Lord thy God hath divided unto all nations under the whole heaven."

This is the legitimate use of the heavenly bodies, just as the worship of them is their abuse, for the division of time—in other words, the formation of a calendar—is a necessity. But as there are many heavenly bodies and several natural divisions of time, the calendars in use by different peoples differ considerably. One division, however, is common to all calendars—the day.

The "day" is the first and shortest natural division of time. At present we recognize three kinds of "days"— *the sidereal day*, which is the interval of time between successive passages of a fixed star over a given meridian; *the apparent solar day*, which is the interval between two passages of the sun's centre over a given meridian, or the interval between two successive noons on a sundial; and *the mean solar day*, which is the interval between the successive passages of a fictitious sun moving uniformly eastward in the celestial equator, and completing its annual course in exactly the same time as that in which the actual sun makes the circuit of the ecliptic. The mean solar days are all exactly the same length; they are equal to the length of the average apparent solar day; and they are each four minutes longer than a sidereal day. We divide our days into 24 hours; each hour into 60 minutes; each minute into 60 seconds. This subdivision of the day requires some mechanical means of continually registering time, and for this purpose we use clocks and watches.

The sidereal day and the mean solar day necessitate some means of registering time, such as clocks; therefore the original day in use must have been the apparent solar day. It must then have been reckoned either from sunset to sunset, or from sunrise to sunrise. Later it might have been possible to reckon it from noon to noon, when some method of fixing the moment of noon had been invented; some method, that is to say, of fixing the true north and south, and of noting that the sun was due south, or the shadow due north. Our own reckoning from midnight to midnight is a late method. Midnight is not marked by the peculiar position of any visible heavenly body; it has, in general, to be registered by some mechanical time-measurer.

In the Old Testament Scriptures the ecclesiastical reckoning was always from one setting of the sun to the next. In the first chapter of Genesis the expressions for the days run, "The evening and the morning," as if the evening took precedence of the morning. When the Passover was instituted as a memorial feast, the command ran—

"In the first month, on the fourteenth day of the month at even, ye shall eat unleavened bread, until the one and twentieth day of the month at even. Seven days shall there be no leaven."

And again, for the sabbath of rest in the seventh month—

"In the ninth day of the month at even, from even unto even, shall ye celebrate your sabbath."

The ecclesiastical "day" of the Jews, therefore, began in the evening, with sunset. It does not by any means follow that their civil day began at this time. It would be more natural for such business contracts as the hiring of servants or labourers to date from morning to morning rather than from evening to evening. Naturally any allusion in the Scriptures to the civil calendar as apart from the ecclesiastical would be indirect, but that common custom was not entirely in agreement with the ecclesiastical formula we may perhaps gather from the fact that in the Old Testament there are twenty-six cases in which the phrases "day and night," "day or night" are employed, and only three where "night" comes before "day." We have a similar divergence of usage in the case of our civil and astronomical days; the first beginning at midnight, and the second at the following noon, since the daylight is the time for work in ordinary business life, but the night for the astronomers. The Babylonians, at least at a late date in their history, had also a twofold way of determining when the day began. Epping and Strassmaier have translated and elucidated a series of Babylonian lunar calendars of dates between the first and second centuries before our era. In one column of these was given the interval of time which elapsed between the true new moon and the first visible crescent.

"Curious to relate, at first all Father Epping's calculations to establish this result were out by a mean interval of six hours. The solution was found in the fact that the Babylonian astronomers were not content with such a variable instant of time as sunset for their calculations, as indeed they ought not to have been, but used as the origin of the astronomical day at Babylon the midnight which followed the setting of the sun, marking the beginning of the civil day."

It may be mentioned that the days as reckoned from sunset to sunset, sunrise to sunrise, and noon to noon, would give intervals of slightly different lengths. This would, however, be imperceptible so long as their lengths were not measured by some accurate mechanical time-measurer such as a clepsydra, sandglass, pendulum, or spring clock.

The first obvious and natural division of the whole day-interval is into the light part and the dark part. As we have seen in Genesis, the evening and the morning are the day. Since Palestine is a sub-tropical country, these would never differ very greatly in length, even at midsummer or midwinter.

The next subdivision, of the light part of the day, is

THE DAY AND ITS DIVISIONS

into morning, noon and evening. As David says in the fifty-fifth Psalm—

"Evening, and morning, and at noon, will I pray."

None of these three subdivisions were marked out definitely in their beginning or their ending, but each contained a definite epoch. Morning contained the moment at which the sun rose; noon the moment at which he was at his greatest height, and was at the same time due south; evening contained the moment at which the sun set.

In the early Scriptures of the Old Testament, the further divisions of the morning and the evening are still natural ones.

For the progress of the morning we have, first, the twilight, as in Job—

"Let the stars of the twilight thereof be dark;
Let it look for light but have none;
Neither let it see the eyelids of the morning."

Then, daybreak, as in the Song of Solomon—

"Until the day break (literally, breathe) and the shadows flee away,"

where the reference is to the cool breezes of twilight. So too in Genesis, in Joshua, in the Judges and in Samuel, we find references to the "break of day" (literally, the rising of the morning, or when it became light to them) and "the dawning of the day" or "about the spring of the day."

The progress of the morning is marked by the increasing heat; thus as " the sun waxed hot," the manna melted; whilst Saul promised to let the men of Jabesh-Gilead have help " by that time the sun be hot," or, as we should put it, about the middle of the morning.

Noon is often mentioned. Ish-bosheth was murdered as he "lay on a bed at noon," and Jezebel's prophets "called on the name of Baal from morning even unto noon."

We find the "afternoon" (lit. " till the day declined ") mentioned in the nineteenth chapter of the Judges, and in the same chapter this period is further described in "The day draweth toward evening (lit. is weak)," and " The day groweth to an end " (lit. " It is the pitching time of the day," that is to say, the time for pitching tents, in preparation for the nightly halt).

As there was no dividing line between the morning and noontide, neither was there any between the afternoon and evening. The shadows of the night were spoken of as chased away by the cool breezes of the morning, so the lengthening shadows cast by the declining sun marked the progress of the evening. Job speaks of the servant who " earnestly desireth the shadow;" that is to say, the intimation, from the length of his own shadow, that his day's work was done; and Jeremiah says, " The shadows of the evening are stretched out." Then came sundown, and the remaining part of the evening is described in Proverbs: "In the twilight, in the evening, in the black and dark night."

In a country like Palestine, near the tropics, with the days not differing extravagantly in length from one part of the year to another, and the sun generally bright and shining, and throwing intense shadows, it was easy, even for the uneducated, to learn to tell the time of day from the length of the shadow. Here, in our northern latitude, the problem is a more complex one, yet we learn from the *Canterbury Tales*, that Englishmen in the time of the Plantagenets could read the position of the sun with quite sufficient accuracy for ordinary purposes. Thus the host of the Tabard inn, though not a learned man—

> "Saw wel, that the brighte sonne
> The ark of his artificial day had ronne
> The fourthe part, and half an houre and more;
> And though he was not depe experte in lore,
> He wiste it was the eighte and twenty day
> Of April, that is messager to May;
> And saw wel that the shadow of every tree
> Was as in lengthe of the same quantitee
> That was the body erect, that caused it;
> And therfore by the shadow he toke his wit,
> That Phebus, which that shone so clere and bright,
> Degrees was five and fourty clombe on hight;
> And for that day, as in that latitude,
> It was ten of the clok, he gan conclude." [1]

In the latter part of the day there is an expression used several times in Exodus, Leviticus, and Numbers "between the two evenings" which has given rise to much controversy. The lamb of the Passover was killed in this period; so also was the lamb of the first year offered daily at the evening sacrifice; and day by day Aaron was then commanded to light the seven lamps and burn incense. It is also mentioned once, in no connection with the evening

[1] *The Man of Lawe's Prologue*, lines 4421-4434.

sacrifice, when the Lord sent quails to the children of Israel saying, "At even (between the two evenings) ye shall eat flesh." In Deuteronomy, where a command is again given concerning the Passover, it is explained that it is "at even, at the going down of the sun." The Samaritans, the Karaite Jews, and Aben Ezra held "the two evenings" to be the interval between the sun's setting and the entrance of total darkness; *i. e.* between about six o'clock and seven or half-past seven. A graphic description of the commencement of the sabbath is given in Disraeli's novel of *Alroy*, and may serve to illustrate this, the original, idea of "between the two evenings."

"The dead were plundered, and thrown into the river, the encampment of the Hebrews completed. Alroy, with his principal officers, visited the wounded, and praised the valiant. The bustle which always succeeds a victory was increased in the present instance by the anxiety of the army to observe with grateful strictness the impending sabbath.

"When the sun set the sabbath was to commence. The undulating horizon rendered it difficult to ascertain the precise moment of his fall. The crimson orb sunk below the purple mountains, the sky was flushed with a rich and rosy glow. Then might be perceived the zealots, proud in their Talmudical lore, holding the skein of white silk in their hands, and announcing the approach of the sabbath by their observation of its shifting tints. While the skein was yet golden, the forge of the armourers still sounded, the fire of the cook still blazed, still the cavalry led their steeds to the river, and still the busy footmen braced up their tents, and hammered at their palisades. The skein of silk became rosy, the armourer worked with renewed energy, the cook puffed with increased zeal, the

THE DAY AND ITS DIVISIONS 279

horsemen scampered from the river, the footmen cast an anxious glance at the fading light.

"The skein of silk became blue ; a dim, dull, sepulchral, leaden tinge fell over its purity. The hum of gnats arose, the bat flew in circling whirls over the tents, horns sounded from all quarters, the sun had set, the sabbath had commenced. The forge was mute, the fire extinguished, the prance of horses and the bustle of men in a moment ceased. A deep, a sudden, an all-pervading stillness dropped over that mighty host. It was night ; the sacred lamps of the sabbath sparkled in every tent of the camp, which vied in silence and in brilliancy with the mute and glowing heavens."

In later times, on account of ritualistic necessities, a different interpretation was held. So Josephus says : "So these high-priests, upon the coming of their feast which is called the Passover, . . . slay their sacrifices, from the ninth hour till the eleventh."[1] And the Talmud made the first evening to begin with the visible decline of the sun and the second with sunset, or " the two evenings " to last from three till about six. Schiaparelli gives the first evening from sunset until the time that the newly visible lunar crescent could be seen in the twilight sky, or about half an hour after sunset, and the second evening from that until darkness set in, basing his argument on the directions to Aaron to light the lamps " between the two evenings," since, he argues, these would not be made to burn in the daylight. Probably in the days of Moses and Aaron the period could not be defined as accurately as this would imply, as the opportunity of seeing the new moon could only come once a month, and we have no

[1] Josephus, *Wars*, VI. ix. 3.

evidence of any mechanical time-measurer being then in use with them.

For shorter spaces of time we have the word "moment" or "instant" many times mentioned. The words may mean, the opening or winking of the eye, "the twinkling of an eye," spoken of by St. Paul, in his Epistle to the Corinthians, and do not describe any actual duration of time, or division of the day.

The only time-measurer mentioned in the Bible is the dial of Ahaz, which will form the subject of a later chapter. It need only be noted here that, as it depended upon the fall of the shadow, it was of use only whilst the sun was shining; not during cloudy weather, or at night.

As the day had three main divisions, so had also the night. There were three "watches," each, like the watches on ship-board, about four hours in length. So in the Psalms, " the watches " are twice put as an equivalent for the night.

The ancient Hebrews would have no difficulty in roughly dividing the night into three equal parts, whenever the stars could be seen. Whether they watched "Arcturus and his sons,"—the circumpolar constellations moving round like a vast dial in the north—or the bringing forth of Mazzaroth, the zodiacal constellations, in the south, they would soon learn to interpret the signs of night with sufficient accuracy for their purpose.

The first watch of night is mentioned in the book of Lamentations.

" Arise, cry out in the night: in the beginning of the watches pour out thine heart like water before the face of the Lord."

THE DAY AND ITS DIVISIONS 281

It was "in the beginning of the middle watch; and they had but newly set the watch," that Gideon and his gallant three hundred made their onslaught on the host of the Midianites.

It was in the third, the morning watch, that "the Lord looked unto the host of the Egyptians through the pillar of fire and of the cloud, and troubled the host of the Egyptians" as they pursued Israel into the midst of the Red Sea. In this watch also, Saul surprised the Ammonites as they besieged Jabesh-Gilead, and scattered them, "so that two of them were not left together."

In the New Testament, the Roman method of dividing the night is adopted; viz. into four watches. When the disciples were crossing the Sea of Galilee in their little boat, and they had toiled all night in rowing because the wind was contrary, it was in "the fourth watch of the night" that Jesus came unto them.

There is no mention of any mechanical time-measurer in the Old Testament, and in only one book is there mention in the English version of the word "hour." Five times it is mentioned in the Book of Daniel as the rendering of the Chaldean word *sha'ah*, which literally means "the instant of time."

No mention either is made of the differing lengths of the days or nights throughout the year—at midsummer the day is $14\frac{1}{4}$ hours long, and the night $9\frac{3}{4}$. Job speaks, however, of causing "the day-spring to know its place," which may well refer to the varying places along the eastern horizon at which the sun rose during the course of the year. Thus in mid-winter the sun rose 28° south of the east point, or half a point south of E.S.E. Similarly in midsummer it rose 28° north of east, or half a point north of E.N.E.[1]

The Babylonians divided the whole day interval into twelve *kasbu*, or "double hours." These again were divided into sixty parts, each equal to two of our minutes; this being about the time that is required for the disc of the sun to rise or set wholly. The Babylonian *kasbu* was not only a division of time, but a division of space, signifying the space that might be marched in a *kasbu* of time. Similarly we find, in the Old Testament, the expression "a day's journey," or "three days' journey," to express distance, and in the New Testament we find the same idea applied to a shorter distance in the "sabbath-day's journey," which was about two miles. But the Jews in New Testament times adopted, not the Babylonian day of twelve hours, but the Egyptian of twenty-four. So we find, in the parable of the Labourers in the Vineyard, mention made of hiring early in the morning, and at the third, sixth, ninth, and eleventh hours; and since those hired latest worked for but one hour, it is evident that there were twelve hours in the daylight. Our Lord alludes to this expressly in the Gospel according to St. John, where he says—

"Are there not twelve hours in the day? If any man walk in the day, he stumbleth not, because he seeth the light of this world. But if a man walk in the night, he stumbleth, because there is no light in him."

[1] See the diagram on p. 363.

CHAPTER II

THE SABBATH AND THE WEEK

THE present chapter has little, if anything, to do with astronomy, for the week, as such, is not an astronomical period. But the sabbath and the week of seven days are so intimately connected with the laws and customs of Israel that it is impossible to leave them out of consideration in dealing with the "times and seasons" referred to in the Bible.

The day, the month and the year are each defined by some specific revolution of one of the great cosmical bodies; there is in each case a return of the earth, or of the earth and moon together, to the same position, relative to the sun, as that held at the beginning of the period.

The week stands in a different category. It is not defined by any astronomical revolution; it is defined by the return of the sabbath, the consecrated day.

A need for the division of time into short periods, less than a month, has been generally felt amongst civilized men. Business of state, commercial arrangements, social intercourse, are all more easily carried out, when some such period is universally recognized. And so, what we may loosely term a "week," has been employed in many ancient nations. The Aztecs, using a short month of 20 days, divided it into four quarters of 5 days each. The Egyptians, using a conventional month of 30 days, divided it into 3 decades; and decades were also used by the Athenians, whose months were alternately of 29 and of 30 days.

Hesiod tells us that the days regarded as sacred in his day were the fourth, fourteenth and twenty-fourth of each month.

> "The fourth and twenty-fourth, no grief should prey
> Within thy breast, for holy either day.
> . . . .
> Pierce on the fourth thy cask; the fourteenth prize
> As holy; and, when morning paints the skies,
> The twenty-fourth is best."

The Babylonians divided the month somewhat differently; the seventh, fourteenth, nineteenth, twenty-first and twenty-eighth days being regarded as "sabbaths."[1]

The sabbath enjoined upon the Hebrews was every seventh day. The week as defined by it was a "free" week; it was tied neither to month nor year, but ran its course uninterruptedly, quite irrespective of the longer divisions of time. It was, therefore, a different conception from that underlying the usages of the Greeks or Babylonians, and, it may be added, a more reasonable and practical one.

Four origins have been assigned for the week. There

---

[1] This is learnt from a single tablet of a Babylonian Calendar (preserved in the British Museum), which unfortunately contains one month only.

THE SABBATH AND THE WEEK 285

are those who assert that it is simply the closest possible approximation to the quarter-month; the mean month being $29\frac{1}{2}$ days in length, a quarter-month would be $7\frac{3}{8}$ days, and since fractions of a day cannot be recognized in any practical division of time for general use, the week of seven days forms the nearest approach to the quarter-month that could be adopted. This is undeniably true, but it is far more likely that such an origin would give rise to the Babylonian system than to the Jewish one, for the Babylonian system corrected the inequality of quarter-month and week every month, and so kept the two in harmony; whilst the Hebrew disregarded the month altogether in the succession of his weeks.

Next, it is asserted that the Hebrew sabbath was derived from the Babylonian, and that "it is scarcely possible for us to doubt that we owe the blessings decreed in the sabbath or Sunday day of rest in the last resort to that ancient and civilized race on the Euphrates and Tigris." [1]

There are two points to be considered here. Did the Babylonians observe their "sabbaths" as days of rest; and, were they or the Hebrews the more likely to hand on their observances to another nation?

We can answer both these questions. As to the first, a large number of Babylonian documents on tablets, preserved in the British Museum, have been published by Father Strassmaier, and discussed by Prof. Schiaparelli. In all there were 2,764 dated documents available for

[1] *Babel and Bible*, Dr. Fried. Delitzsch, Johns' Translation, pp. 40, 41.

examination, nearly all of them commercial and civil deeds, and covering practically the whole period from the accession of Nebuchadnezzar to the twenty-third year of Darius Hystaspes. This number would give an average of 94 deeds for each day of the month; the number actually found for the four "sabbaths," *i.e.* for the 7th, 14th, 21st and 28th days, were 100, 98, 121 and 91 respectively. The Babylonians evidently did not keep these days as days of rest, or of abstinence from business, as the Jews keep their sabbath, or Christian countries their Sunday. They cannot even have regarded it as an unlucky day, since we find the average of contracts is rather higher for a "sabbath" than for a common day.

The case is a little different with the 19th day of the month. This, as the 49th day from the beginning of the previous month, was a sabbath of sabbaths, at the end of a "week of weeks." In this case only 89 contracts are found, which is slightly below the average, though twelve common days show a lower record still. But in most cases the date is written, not as 19, but as 20-1; as if there were a superstition about the number 19. On the other hand, this method of indicating the number may be nothing more than a mode of writing; just as in our Roman numerals, XIX., one less than XX., is written for 19.

The Babylonians, therefore, did not observe these days as days of rest, though they seem to have marked them in the ritual of temple and court. Nor did they make every seventh or every fifth a rest-day, for Prof. Schiaparelli has specially examined these documents to see if they gave

THE SABBATH AND THE WEEK 287

any evidence of abstention from business either on one day in seven or on one day in five, and in both cases with a purely negative result.

When we inquire which nation has been successful in impressing their particular form of sabbath on the nations around the case is clear. We have no evidence of the Babylonians securing the adoption of their sabbatic arrangements by the Persians, Greeks and Parthians who successively overcame them. It was entirely different with the Jews. The Jewish kingdom before the Captivity was a very small one compared with its enemies on either side—Assyria, Babylon and Egypt; it was but a shadow even of its former self after the Return. And imperial Rome was a mightier power than Assyria or Babylon at their greatest. If ever one state was secure from influence by another on the score of its greater magnitude and power, Rome was safe from any Jewish impress. Yet it is perfectly well known that the impression made upon the Romans by the Jews in this very matter of sabbath-keeping was widespread and deep. Jewish influence was felt and acknowledged almost from the time that Syria, of which Judæa was but a petty division, became a Roman province, and a generation had not passed away before we find Horace making jocular allusion to the spread of the recognition of the Jewish sabbath. In his ninth satire he describes himself as being buttonholed by a bore, and, seeing a friend pass by, as begging the latter to pretend business with him and so relieve him of his trouble. His friend mischievously excuses himself from talking about business:—

> "To-day's the thirtieth sabbath. Can you mean
> Thus to insult the circumcised Jews?"

Persius, in his fifth satire, speaks of those who—

> "Move their lips with silence, and with fear
> The sabbath of the circumcised revere."

Juvenal, in his fourteenth satire, describes how many Romans reverence the sabbath; and their sons, bettering the example, turn Jews themselves:—

> "Others there are, whose sire the sabbath heeds,
> And so they worship naught but clouds and sky.
> They deem swine's flesh, from which their father kept,
> No different from a man's. And soon indeed
> Are circumcised; affecting to despise
> The laws of Rome, they study, keep and fear
> The Jewish law, whate'er in mystic book
> Moses has handed down,—to show the way
> To none but he who the same rites observes,
> And those athirst to lead unto the spring
> Only if circumcised. Whereof the cause
> Was he, their sire, to whom each seventh day
> Was one of sloth, whereon he took in hand
> No part in life."

Ovid, Tibullus, and others also speak of the Jewish sabbath, not merely as universally known, but as largely observed amongst the Romans, so that it obtained almost a public recognition, whilst the success of Judaism in making proselytes, until Christianity came into rivalry with it, is known to every one.

As to the general influence of Judaism in securing the recognition of the week with its seventh day of rest, the testimony of Josephus is emphatic.

THE SABBATH AND THE WEEK 289

"The multitude of mankind itself have had a great inclination of a long time to follow our religious observances; for there is not any city of the Grecians, nor any of the barbarians, nor any nation whatsoever, whither our custom of resting on the seventh day hath not come, and by which our fasts and lighting up lamps, and many of our prohibitions as to our food, are not observed; they also endeavour to imitate our mutual concord with one another, and the charitable distribution of our goods, and our diligence in our trades, and our fortitude in undergoing the distresses we are in, on account of our laws; and, what is here matter of the greatest admiration, our law hath no bait of pleasure to allure men to it, but it prevails by its own force; and as God Himself pervades all the world, so hath our law passed through all the world also."[1]

Philo, the Jew, bears equally distinct testimony to the fact that wheresoever the Jews were carried in their dispersion, their laws and religious customs, especially their observance of every seventh day, attracted attention, and even secured a certain amount of acceptance. The Jews, therefore, even when, as a nation, they were ruined and crushed, proved themselves possessed of such vital force, of such tenacity, as to impress their conquerors with interest in, and respect for, their sabbatic customs. Of their tenacity and force in general, of their power to influence the nations amongst whom they have been scattered, the history of the last two thousand five hundred years is eloquent. It is not reasonable, nor scientific, to suppose that this nation, steel since it returned from its captivity in Babylon, was wax before.

But the third suggestion as to the origin of the week

[1] *Flavius Josephus against Apion*, book ii. 40.

of seven days,—that it was derived from the influence of the planets,—makes the matter clearer still. This suggestion has already been noticed in the chapter on "Saturn and Astrology." It is sufficient to say here that it presupposes a state of astronomical advancement not attained until long after the sabbath was fully known. The Babylonians did observe the seven planets, but there is no trace of their connection with the Babylonian week. But when the Greek astronomers had worked out that system of the planetary motions which we call after Ptolemy, and the planets had been fitted by the Alexandrian observers to the days of the Jewish week and the hours of the Egyptian day, then the Babylonian astrologers also adopted the mongrel combination. Thus indirectly Babylon received the free week from the Jews, and did not give it.

"The oldest use of the free and uniform week is found among the Jews, who had only a most imperfect knowledge of the planets. The identity of the number of the days in the week with that of the planets is purely accidental, and it is not permissible to assert that the former number is derived from the latter."[1]

"Carried by the Jews into their dispersion, adopted by the Chaldæan astrologers for use in their divinations, received by Christianity and Islam, this cycle" (the free week of seven days), "so convenient and so useful for chronology, has now been adopted throughout the world. Its use can be traced back for about 3,000 years, and there is every reason to believe that it will last through the centuries to come, resisting the madness of useless novelty and the assaults of present and future iconoclasts."[2]

[1] Schiaparelli, *Astronomy in the Old Testament*, p. 135.
[2] *Ibid.*, p. 133.

THE SABBATH AND THE WEEK 291

The fourth account of the origin of the week is that given us in the Bible itself.

" In six days the Lord made heaven and earth, the sea, and all that in them is, and rested the seventh day: wherefore the Lord blessed the sabbath day, and hallowed it."

The institution of the sabbath day is the crown of the work of creation, the key to its purpose. Other times and seasons are marked out by the revolutions and conjunctions of the heavenly bodies. This day is set apart directly by God Himself; it is His express handiwork,— " the day which the Lord hath made."

The great truth taught in the first chapter of Genesis is that God is the One Reality. All that we can see above or around was made by Him. He alone is God.

And His creative work has a definite goal to which its several details all lead up—the creation of man, made in the image of God.

As such, man has a higher calling than that of the beasts that perish. The chief object of their lives is to secure their food; their aspirations extend no further. But he is different; he has higher wants, nobler aspirations. How can they be met?

The earth was created to form an abode suitable for man; the varied forms of organic life were brought into existence to prepare the way for and minister to him For what was man himself made, and made in the image of God, but that he might know God and have communion with Him? To this the sabbath day gave the call, and for this it offered the opportunity.

> "For what are men better than sheep or goats,
> That nourish a blind life within the brain,
> If, knowing God, they lift not hands of prayer?"

CHAPTER III

THE MONTH

THE shortest natural division of time is the day. Next in length comes the month.

As was pointed out in the chapter on the Moon, the Hebrews used two expressions for month—*Chodesh*, from a root meaning "to be new"; and *Yerach*, from the root meaning "to be pale."

Chodesh is the word most commonly employed, and this, in itself, is sufficient to show that the Hebrew calendar month was a lunar one. But there are, besides, too many references to the actual new moons for there to be any doubt on the question.

Every seventh day was commanded to be held as a sabbath of rest, and on it were sacrificed four lambs, instead of the two offered up, the one at the morning and the other at the evening sacrifice of the six working days. But the new moons are also mentioned as holy days, and are coupled with the sabbaths. The husband of the Shunamite asked her why she wished to go to Elisha, as "it is neither new moon, nor sabbath." Isaiah, speaking in the name of the Lord, says—

"The new moons and sabbaths, the calling of assemblies,

I cannot away with; ... your new moons and your appointed feasts My soul hateth"; and again, "From one new moon to another, and from one sabbath to another, shall all flesh come to worship."

Amos speaks of degenerate Israel, that they say—

"When will the new moon be gone, that we may sell corn? and the sabbath, that we may set forth wheat?"

As late as Apostolic times, St. Paul refers to the feasts of the new moons, saying, "Let no man therefore judge you ... in respect ... of the new moon."

The ordinances respecting the observance of the new moons—the "beginnings of months"—were explicit. Trumpets were blown over the burnt offerings and over the sacrifices of the peace offerings, and the nature of these offerings is given in detail in the twenty-eighth chapter of the Book of Numbers. The ordinances were reiterated and emphasized in the days of David, Solomon, Hezekiah Ezekiel, Ezra and Nehemiah. Amongst the Jews of the present day the trumpets are not blown at new moons; extra prayers are read, but the burnt and peace offerings are of necessity omitted.

Beside the "new moons" and the sabbaths, the ancient Hebrews had three great festivals, all defined as to the time of their celebration by the natural months.

The first was the Feast of the Passover, which lasted a week, and began with the killing of a lamb "between the two evenings"; on the 14th day of the month Abib, the first month of the year—that is to say, on the evening that the first moon of the year became full. This feast corresponded to our Easter. The second was that of Pentecost, and was bound to the Feast of the Passover by being appointed to occur seven weeks after the consecration of the harvest season by the offering of the sheaf on the second day of the Passover. We still celebrate the Feast of Pentecost, or Whitsunday, keeping it in remembrance of the birthday of the Christian Church. This feast lasted but a single day, and did not occur at either the new or the full of the moon, but nearly at first quarter.

The third festival was threefold in its character. It began with special sacrifices besides those usually offered at the new moon:—

"In the seventh month, on the first day of the month, ye shall have an holy convocation; ye shall do no servile work: it is a day of blowing of trumpets unto you."

This then was especially dependent on the new moon, being on the first day of the month.

On the 10th day of the month was the Day of Atonement, when the people should afflict their souls. On the 15th day of the month began the Feast of Tabernacles, which commenced on the night that the moon was full, and lasted for a week.

We have no special religious seasons in the Christian Church to correspond with these.

We thus see that with the Hebrews all the days of the new moons, and two days of full moon (in the first and in the seventh months), were days for which special ordinances were imposed. And there is no doubt that the beginnings of the new months were obtained by direct observation of the moon, when weather or other conditions permitted, not by any rule of thumb computation. The new moon observed was, necessarily, not the new moon as understood in the technical language of astronomy; *i.e.* the moment when the moon is in "conjunction" with the sun, having its dark side wholly turned towards the earth, and being in consequence completely invisible. "The new moon" as mentioned in the Scriptures, and as we ordinarily use the term, is not this conjunction, but the first visible crescent of the moon when it has drawn away from the sun sufficiently to be seen after sunset for a short time, in the twilight, before it sets; for the moon when very slender cannot be seen in daylight. It may, therefore, be first seen any time between about 18 hours and 40 hours after its conjunction with the sun; in other words, it may be first seen on one of two evenings. But for the ecclesiastical rites it was necessary that there should be an authoritative declaration as to the time of the commencement of the month, and, moreover, the great feasts were fixed for certain days in the month, and so were dependent on its beginning.

During the period of the Jewish restoration, up to the destruction of Jerusalem by Titus, the Sanhedrim used to sit in the "Hall of Polished Stones" to receive the testimony of credible witnesses that they had seen the new moon. If the new moon had appeared at the commencement of the 30th day—corresponding to our evening of the 29th—the Sanhedrim declared the previous month "imperfect," or consisting only of 29 days.

If credible witnesses had not appeared to testify to the appearance of the new moon on the evening of the 29th, the next evening, *i.e.* that of the 30th—according to our mode of reckoning—was taken as the commencement of the new month, and the previous month was then declared to be "full," or of 30 days.

Early in the Christian era, it was enacted that no testimony should be received from unknown persons, because, says the Talmud, the Baithusites wished to impose on the Mishnic Rabbis, and hired two men to do so for four hundred pieces of silver.

It is clear, therefore, that about the time of the Christian era the beginnings of the months were determined astronomically from the actual observation of the new moons, and we may safely conclude that it was the same also from the earliest times. It was the actual new moon, not any theoretical or fictitious new moon, that regulated the great festivals, and, as we have seen, there was often some considerable uncertainty possible in the fixing of the dates. The witnesses might give conflicting testimony, and the authoritative date might be proved to be in fault. We have an instance of such conflicting authority in the different dating, on one occasion, of the Day of Atonement by the Rabbi Yehoshua, and Rabbon Gamaliel, the president of the Sanhedrim, grandson of the Gamaliel at whose feet Paul sat.

According to a statement in the Mishna, dating from the second century of our era, the appearance of the new moon at Jerusalem was signalled to Babylonia during the century preceding the destruction of the Holy City by

Titus, and perhaps from earlier times. The dispersion of the Jews had therefore presented them with an additional difficulty in fixing the beginning of their months. The problem is much more intricate to-day, seeing that the Jews are dispersed over the whole world, and the new moon, first visible on one evening at Jerusalem, might be seen the evening before, according to the reckoning of places west of Jerusalem, or might be invisible until the following evening, according to the reckoning of places east of it. We have the same problem to solve in finding the date of Easter Sunday. The Prayer Book rule for finding it runs thus:—

"Easter day is always the first Sunday after the full moon which happens upon, or next after, the 21st day of March; and if the full moon happens on a Sunday, Easter day is the Sunday after."

But the "moon" we choose for the ecclesiastical calendar is an imaginary body, which is so controlled by specially constructed tables as to be "full" on a day not differing by more than two or three days at most from the date on which the actual moon is full. This may seem, at first sight, a very clumsy arrangement, but it has the advantage of defining the date of Easter precisely, without introducing any question as to the special meridian where the moon might be supposed to be observed. Thus, in 1905, the moon was full at $4^h\ 56^m$ Greenwich mean time on the morning of March 21. But Easter Day was not fixed for March 26, the next Sunday following that full moon, but a month later, for April 23. For the calendar moon, the imaginary moon, was full on March 20; and it may be added that the actual moon, though full on March 21 for European time, was full on March 20 for American time. There would have been an ambiguity, therefore, if the actual moon had been taken, according to the country in which it was observed, an ambiguity which is got rid of by adopting a technical or imaginary moon.

The names given to the different months in Scripture have an interest of their own. For the most part the months are simply numbered; the month of the Passover is the first month, and the others follow, as the second, third, fourth, etc., throughout the year; examples of each occurring right up to the twelfth month. There is no mention of a thirteenth month.

But occasionally we find names as well as numbers given to the months. The first of these is Abib, meaning the month of "green ears." This was the first month, the month of the Passover, and it received its name no doubt from the first green ears of barley offered before the Lord during the feast that followed the Passover.

The second month was called Zif, "splendour"; apparently referring to the splendour of the flowers in full spring time. It is mentioned together with two other names, Ethanim, the seventh month, and Bul, the eighth month, in the account of the building and dedication of Solomon's Temple. The last two are certainly Phœnician names, having been found on Phœnician inscriptions; the first is possibly Phœnician also. Their occurrence in this special connection was no doubt a result of the very large part taken in the building of the Temple and the construction of its furniture by the workmen of Hiram, king of Tyre. The Phœnician names of the months would naturally appear in the contracts and accounts for the work, side by side with the Hebrew equivalents; just as an English contractor to-day, in negotiating for a piece of work to be carried out in Russia, would probably take care to use the dating both of the Russian old style calendar, and of the English new style. The word used for month in these cases is generally, not *chodesh*, the month as beginning with the new moon, but *yerach*, as if the chronicler did not wish them to be understood as having been determined by Jewish authorities or methods. In one case, however, *chodesh* is used in connection with the month Zif.

The other instances of names for the months are Nisan, Sivan, Elul, Chisleu, Tebeth, Sebat, and Adar, derived from month names in use in Babylonia, and employed only in the books of Esther, Ezra, Nehemiah, and Zechariah, all avowedly post-exilic writers. The month word used in connection with them is *chodesh*—since the Babylonian months were also lunar—except in the single case where Ezra used a month name, terming it *yerach*. The other post-exilic writers or editors of the books of Holy Scripture would seem to have been at some pains to omit all Babylonian month names. These Babylonian month names continue to be used in the Jewish calendar of to-day.

In four places in Scripture mention is made of a month of days, the word for month being in two cases *chodesh*, and in two, *yerach*. Jacob, when he came to Padan-aram, abode with Laban for "the space of a month," before his crafty uncle broached the subject of his wages. This may either merely mean full thirty days, or the term *chodesh* may possibly have a special appropriateness, as Laban may have dated Jacob's service so as to commence from the second new moon after his arrival. Again, when the people lusted for flesh in the wilderness, saying, "Who shall give us flesh to eat?" the Lord promised to send them flesh—

"And ye shall eat. Ye shall not eat one day, nor two days, nor five days, neither ten days, nor twenty days, but even a whole month. . . . And there went forth a wind from the Lord, and brought quails from the sea."

> "He rained flesh also upon them as dust,
> And feathered fowls like as the sand of the sea."

The "whole month" in this case was evidently a full period of thirty days, irrespective of the particular phase of the moon when it began and ended.

Amongst the Babylonians the sign for the word month was xxx, expressing the usual number of days that it contained, and without doubt amongst the Hebrews that was the number of days originally assigned to the month, except when the interval between two actually observed new moons was found to be twenty-nine. In later times it was learned that the length for the lunation lay between twenty-nine and thirty days, and that these lengths for the month must be alternate as a general rule. But in early times, if a long spell of bad weather prevented direct observation of the new moon, we cannot suppose that anything less than thirty days would be assigned to each month.

Such a long spell of bad observing weather did certainly occur on one occasion in the very early days of astronomy, and we accordingly find that such was the number of days allotted to several consecutive months, though the historian was evidently in the habit of observing the new moon, for *chodesh* is the word used to express these months of thirty days.

We are told that—

"In the six hundreth year of Noah's life, in the second month, the seventeenth day of the month, the same day were all the fountains of the great deep broken up, and the windows of heaven were opened."

And later that—

"After the end of the hundred and fifty days the waters were abated. And the ark rested in the seventh month, on the seventeenth day of the month, upon the mountains of Ararat."

The five months during which the waters prevailed upon the earth were, therefore, reckoned as of thirty days each. If all the new moons, or even that of the seventh month, had been actually observed, this event would have been ascribed to the nineteenth day of the month, since 150 days is five months and two days; but in the absence of such observations a sort of "dead reckoning" was applied, which would of course be corrected directly the return of clear weather gave an opportunity for observing the new moon once again.

A similar practice was followed at a much later date in Babylon, where astronomy is supposed to have been highly developed from remote antiquity. Thus an inscription recently published by Dr. L. W. King records that—

"On the 26th day of the month Sivan, in the seventh year, the day was turned into night, and fire in the midst of heaven."

This has been identified by Mr. P. H. Cowell, F.R.S., Chief Assistant at the Royal Observatory, Greenwich, as the eclipse of the sun that was total at Babylon on July 31, B.C. 1063. The Babylonians, when bad weather obliged them to resort to dead reckoning, were, therefore, still reckoning the month as precisely thirty days so late as the times of Samuel and Saul, and in this particular instance were two, if not three, days out in their count. Had the new moon of Sivan been observed, or correctly calculated, the eclipse must have been reckoned as falling on the 28th or 29th day of the month.

The Athenians in the days of Solon, five hundred years later than this, adopted months alternately twenty-nine and thirty days in length, which gives a result very nearly correct.

The Jews after the Dispersion adopted the system of thus alternating the lengths of their months, and with some slight modifications it holds good to the present day. As will be shown in the following chapter, the ordinary years are of twelve months, but seven years in every nineteen are "embolismic," having an extra month. The names employed are those learned during the Babylonian captivity, and the year begins with the month Tishri, corresponding to September-October of our calendar. The lengths of most of the months are fixed as given in the following table, but any adjustment necessary can be effected either by adding one day to Heshvan, which has usually twenty-nine days, or taking away one day from Kislev, which has usually thirty—

	Ordinary Year Days	Embolismic Year Days
Tishri	30	30
Heshvan	29 +	29 +
Kislev	30 −	30 −
Tebeth	29	29
Shebat	30	30
Adar	29	30
Ve-adar	...	29
Nisan	30	30
Yiar	29	29
Sivan	30	30
Tamuz	29	29
Ab	30	30
Elul	29	29

The Jewish month, therefore, continues to be essentially a true lunar one, though the exact definition of each month is, to some extent, conventional, and the words of the Son of Sirach still apply to the Hebrew calendar—

> "The moon also is in all things for her season,
> For a declaration of times, and a sign of the world.
> From the moon is the sign of the feast day;
> A light that waneth when she is come to the full."

For so God—

> "Appointed the moon for seasons."

CHAPTER IV

THE YEAR

THE third great natural division of time is the year, and, like the day and the month, it is defined by the relative apparent movements of the heavenly bodies.

As the Rabbi Aben Ezra pointed out, *shanah*, the ordinary Hebrew word used for year, expresses the idea of *annus* or *annulus*, a closed ring, and therefore implies that the year is a complete solar one. A year, that is purely lunar, consists of twelve lunations, amounting to 354 days. Such is the year that the Mohammedans use; and since it falls short of a solar year of 365 days by 10 or 11 days, its beginning moves backwards rather rapidly through the seasons.

The Jews used actual lunations for their months, but their year was one depending on the position of the sun, and their calendar was therefore a luni-solar one. But lunations cannot be made to fit in exactly into a solar year—12 lunations are some 11 days short of one year; 37 lunations are 2 or 3 days too long for three years—but an approximation can be made by giving an extra month to every third year; or more nearly still by taking 7 years in every 19 as years of 13 months each. This thirteenth month is called an intercalary month, and in the present Jewish calendar it is the month Adar which is reduplicated under the name of Ve-Adar. But, though from the necessity of the case, this intercalation, from time to time, of a thirteenth month must have been made regularly from the first institution of the feast of unleavened bread, we find no allusion, direct or indirect, in the Hebrew Scriptures to any such custom.

Amongst the Babylonians a year and a month were termed "full" when they contained 13 months and 30 days respectively, and "normal" or "incomplete" when they contained but 12 months or 29 days. The succession of full and normal years recurred in the same order, at intervals of nineteen years. For 19 years contain 6939 days $14\frac{1}{2}$ hours; and 235 months, 6939 days $16\frac{1}{2}$ hours; the two therefore differing only by about a couple of hours. The discovery of this cycle is attributed to Meton, about 433 B.C., and it is therefore known as the Metonic cycle. It supplies the "Golden Numbers" of the introduction to the Book of Common Prayer.

There are two kinds of solar years, with which we may have to do in a luni-solar calendar—the tropical or equinoctial year, and the sidereal year. The tropical year is the interval from one season till the return of that season again—spring to spring, summer to summer, autumn to autumn, or winter to winter. It is defined as the time included between two successive passages of the sun through the vernal equinox, hence it is also called the equinoctial year. Its length is found to be 365 days, 5 hours, 49 minutes, and some ancient astronomers derived

THE YEAR

its length as closely as 365 days, 6 hours, by observing the dates when the sun set at exactly the opposite part of the horizon to that where it rose.

The sidereal year is the time occupied by the sun in apparently completing the circuit of the heavens from a given star to the same star again. The length of the sidereal year is 365 days, 6 hours, 9 minutes. In some cases the ancients took the sidereal year from the "heliacal" risings or settings of stars, that is from the interval between the time when a bright star was first seen in the morning just before the sun rose, until it was first so seen again; or last seen just after the sun set in the evening, until it was last so seen again.

But to connect the spring new moon with the day when the sun has returned to the equinox is a more difficult and complicated matter. The early Hebrews would seem to have solved the problem practically, by simply watching the progress of the growing grain. If at one new moon in spring time it appeared clear that some of the barley would be ready in a fortnight for the offering of the green ears at the feast of unleavened bread, then that was taken as beginning the new year. If it appeared doubtful if it would be ready, or certain that it would not be, then the next new moon was waited for. This method was sufficient in primitive times, and so long as the nation of Israel remained in its own land. In the long run, it gave an accurate value for the mean tropical year, and avoided all the astronomical difficulties of the question. It shows the early Hebrews as practical men, for the solution adopted was easy, simple and efficient. This practical method of determining the beginning of the year amongst the early Hebrews, does not appear to have been the one in use amongst the Babylonians either early or late in their history. The early Babylonians used a sidereal year, as will be shown shortly. The later Babylonians used a tropical year dependent on the actual observation of the spring equinox.

To those who have no clocks, no telescopes, no sundials, no instruments of any kind, there are two natural epochs at which the day might begin; at sunrise, the beginning of daylight; and at sunset, the beginning of darkness. Similarly, to all nations which use the tropical year, whether their calendar is dependent on the sun alone, or on both sun and moon, there are two natural epochs at which the year may begin; at the spring equinox, the beginning of the bright half of the year, when the sun is high in the heavens, and all nature is reviving under its heat and light; and at the autumn equinox, the beginning of the dark half of the year, when the sun is low in the heavens, and all nature seems dying. As a nation becomes more highly equipped, both in the means of observing, and in knowledge, it may not retain either of these epochs as the actual beginning of its year, but the determination of the year still rests directly or indirectly upon the observation of the equinoxes.

At the exodus from Egypt, in the month Abib, the children of Israel were commanded in these words—

"This month shall be unto you the beginning of months: it shall be the first month of the year to you."

THE YEAR

This command may have abolished and reversed the previously existing calendar, or it may have related solely to the ecclesiastical calendar, and the civil calendar may have been still retained with a different epoch of commencement.

An inquiry into the question as to whether there is evidence in Scripture of the use of a double calendar, shows that in every case that the Passover is mentioned it is as being kept in the first month, except when Hezekiah availed himself of the regulation which permitted its being kept in the second month. Since the Passover was a spring feast, this links the beginning of the year to the spring time. Similarly the feast of Tabernacles, which is an autumn festival, is always mentioned as being held in the seventh month.

These feasts would naturally be referred to the ecclesiastical calendar. But the slight evidences given in the civil history point the same way. Thus some men joined David at Ziklag during the time of his persecution by Saul, "in the first month." This was spring time, for it is added that Jordan had overflowed all its banks. Similarly, the ninth month fell in the winter: for it was as he "sat in the winter-house in the ninth month, and there was a fire on the hearth burning before him" that king Jehoiakim took the prophecy of Jeremiah and "cut it with the penknife, and cast it into the fire that was on the hearth." The same ninth month is also mentioned in the Book of Ezra as a winter month, a time of great rain.

The same result is given by the instances in which a Babylonian month name is interpreted by its corresponding

Jewish month number. In each case the Jewish year is reckoned as beginning with Nisan, the month of the spring equinox.

In one case, however, two Babylonian month names do present a difficulty.

In the Book of Nehemiah, in the first chapter, the writer says—

"It came to pass in the month Chisleu, in the twentieth year, as I was in Shushan the palace, that Hanani, one of my brethren, came"—

and told him concerning the sad state of Jerusalem. In consequence of this he subsequently approached the king on the subject "in the month Nisan, in the twentieth year of Artaxerxes the king."

If the twentieth year of king Artaxerxes began in the spring, Nisan, which is a spring month, could not follow Chisleu, which is a month of late autumn. But Artaxerxes may have dated his accession, and therefore his regnal years, from some month between Nisan and Chisleu; or the civil year may have been reckoned at the court of Shushan as beginning with Tishri. It may be noted that Nehemiah does not define either of these months in terms of the Jewish. Elsewhere, when referring to the Jewish Feast of Tabernacles, he attributes it to the seventh month, in accord with its place in the Mosaic calendar. An alteration of the beginning of the year from the spring to the autumn was brought about amongst the Jews at a later date, and was systematized in the Religious Calendar by the Rabbis of about the fourth century A.D. Tishri begins the Jewish year at the present day; the first day of Tishri being taken as the anniversary of the creation of the world.

The Mishna, "The Law of the Lip," was first committed to writing in 191 A.D., and the compilation of the Babylonian Talmud, based on the Mishna, was completed about 500 A.D. In its commentary on the first chapter of Genesis, there is an allusion to the year as beginning in spring, for it says that—

"A king crowned on the twenty-ninth of Adar is considered as having completed the first year of his reign on the first of Nisan" (*i.e.* the next day). "Hence follows (observes some one) that the first of Nisan is the new year's day of kings, and that if one had reigned only one day in a year, it is considered as a whole year." [1]

It is not indicated whether this rule held good for the kings of Persia, as well as for those of Israel. If so, and this tradition be correct, then we cannot explain Nehemiah's reckoning by supposing that he was counting from the month of the accession of Artaxerxes, and must assume that a civil or court year beginning with Tishri, *i.e.* in the autumn, was the one in question.

A further, but, as it would seem, quite an imaginary difficulty, has been raised because the feast of ingathering, or Tabernacles, though held in the seventh month, is twice spoken of as being "in the end of the year," or, as it is rendered in the margin in one case, "in the revolution of the year." This latter expression occurs again in 2 Chron. xxiv. 23, when it is said that, "at the end of the

[1] P. I. Hershon, *Genesis with a Talmudical Commentary*, p. 30.

year, the host of Syria came up"; but in this case it probably means early spring, for it is only of late centuries that war has been waged in the winter months. Down to the Middle Ages, the armies always went into winter quarters, and in the spring the kings led them out again to battle. One Hebrew expression used in Scripture means the return of the year, as applied to the close of one and the opening of another year. This is the expression employed in the Second Book of Samuel, and of the First Book of Chronicles, where it is said "after the year was expired, at the time when kings go forth to battle," implying that in the time of David the year began in the spring. The same expression, no doubt in reference to the same time of the year, is also used in connection with the warlike expeditions of Benhadad, king of Syria, and of Nebuchadnezzar, king of Babylon.

It is admitted that the Feast of Tabernacles was held in the autumn, and in the seventh month. The difficulty lies in the question of how it could be said to be "in the end of the year," "at the year's end," although it is clear from the cases just cited that these and similar expressions are merely of a general character, as we ourselves might say, "when the year came round," and do not indicate any rigid connection with a specific date of the calendar.

We ourselves use several years and calendars, without any confusion. The civil year begins, at midnight, on January 1; the financial year on April 1; the ecclesiastical year with Advent, about December 1; the scholastic year about the middle of September, and so on. As

THE YEAR

the word "year" expresses with ourselves many different usages, there is no reason to attribute to the Jews the extreme pedantry of invariably using nothing but precise definitions drawn from their ecclesiastical calendar.

The services of the Tabernacle and the Temple were —with the exception of the slaying of the Paschal lambs—all comprised within the hours of daylight; there was no offering before the morning sacrifice, none after the evening sacrifice. So, too, the Mosaic law directed all the great feasts to be held in the summer half of the year, the light half; none in the winter. The Paschal full moon was just after the spring equinox; the harvest moon of the Feast of Tabernacles as near as possible to the autumn equinox. Until the introduction, after the Captivity, of the Feast of Purim in the twelfth month, the month Adar, the ecclesiastical year might be said to end with those seven days of joyous "camping-out" in the booths built of the green boughs; just as all the great days of the Christian year lie between Advent and the octave of Pentecost, whilst the "Sundays after Trinity" stretch their length through six whole months. There is, therefore, no contradiction between the command in Exod. xii., to make Abib, the month of the Passover, the first month, and the references elsewhere in Exodus to the Feast of Ingathering as being in "the end of the year." It was at the end of the agricultural year; it was also at the end of the period of feasts. So, if a workman is engaged for a day's work, he comes in the morning, and goes home in the evening, and expects to be paid as he leaves; no one would ask him

314   THE ASTRONOMY OF THE BIBLE

to complete the twenty-four hours before payment and dismissal. It is the end of his day; though, like the men in the parable of the Labourers in the Vineyard, he has only worked twelve hours out of the twenty-four. In the same way the Feast of Tabernacles, though in the seventh month, was in "the end of the year," both from the point of view of the farmer and of the ordinances of the sacred festivals.

The method employed in very early times in Assyria and Babylonia for determining the first month of the year was a simple and effective one, the principle of which may be explained thus: If we watch for the appearance of the new moon in spring time, and, as we see it setting in the west, notice some bright star near it, then 12 months later we should see the two together again; but with this difference, that the moon and star would be seen together, not on the first, but on the second evening of the month. For since 12 lunar months fall short of a solar year by 11 days, the moon on the first evening would be about 11 degrees short of her former position. But as she moves about 13 degrees in 24 hours, the next evening she would practically be back in her old place. In the second year, therefore, moon and star would set together on the second evening of the first month; and in like manner they would set together on the third evening in the third year; and, roughly speaking, on the fourth evening of the fourth year. But this last conjunction would mean that they would also set together on the first evening of the next month, which would thus be indicated as the true first month of the year. Thus when moon and star set together on the third evening of a month, thirteen months later they would set together on the first evening of a month. Thus the setting together of moon and star would not only mark which was to be first month of the year, but if they set together on the first evening it would show that the year then beginning was to be an ordinary one of 12 months; if on the third evening, that the year ought to be a full one of 13 months.

This was precisely the method followed by the Akkadians some 4000 years ago. For Prof. Sayce and Mr. Bosanquet translate an old tablet in Akkadian as follows:—

"When on the first day of the month *Nisan* the star of stars (or *Dilgan*) and the moon are parallel, that year is normal. When on the third day of the month *Nisan* the star of stars and the moon are parallel, that year is full."[1]

The "star of stars" of this inscription is no doubt the bright star Capella, and the year thus determined by the setting together of the moon and Capella would begin on the average with the spring equinox about 2000 B.C.

When Capella thus marked the first month of the year, the "twin stars," Castor and Pollux, marked the second month of the year in just the same way. A reminiscence of this circumstance is found in the signs for the first two months; that for the first month being a crescent moon "lying on its back;" that for the second month a pair of stars.

[1] *Monthly Notices of the Royal Astronomical Society*, vol. xxxix. p. 455.

The significance of the crescent being shown as lying on its back is seen at once when it is remembered that the new moon is differently inclined to the horizon according to the time of the year when it is seen. It is most nearly upright at the time of the autumn equinox;

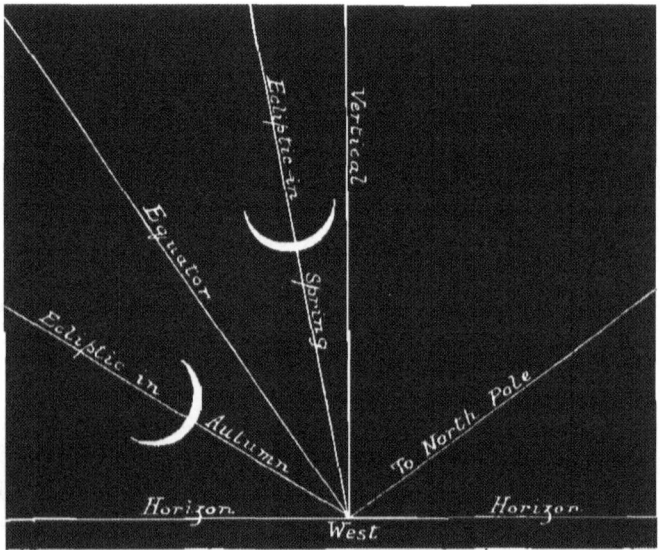

POSITION OF THE NEW MOON AT THE EQUINOXES.

it is most nearly horizontal, "lying on its back," at the spring equinox. It is clear from this symbol, therefore, that the Babylonians began their year in the spring.

This method, by which the new moon was used as a kind of pointer for determining the return of the sun to the neighbourhood of a particular star at the end of a solar year, is quite unlike anything that commentators

BOUNDARY-STONE IN THE LOUVRE; APPROXIMATE DATE, B.C. 1200.
(From a photograph by Messrs. W. A. Mansell.)

on the astronomical methods of the ancients have supposed them to have used. But we know from the ancient inscription already quoted that it was actually used; it was eminently simple; it was bound to have suggested itself wherever a luni-solar year, starting from the observed new moon, was used. Further, it required no instruments or star-maps; it did not even require a knowledge of the constellations; only of one or two conspicuous stars. Though rough, it was perfectly efficient, and would give the mean length of the year with all the accuracy that was then required.

But it had one drawback, which the ancients could not have been expected to foresee. The effect of "precession," alluded to in the chapter on "The Origin of the Constellations," p. 158, would be to throw the beginning of the year, as thus determined, gradually later and later in the seasons,—roughly speaking, by a day in every seventy years,—and the time came, no doubt, when it was noticed that the terrestrial seasons no longer bore their traditional relation to the year. This probably happened at some time in the seventh or eighth centuries before our era, and was connected with the astronomical revolution that has been alluded to before; when the ecliptic was divided into twelve equal divisions, not associated with the actual stars, the Signs were substituted for the Constellations of the Zodiac, and the Ram was taken as the leader instead of the Bull. The equinox was then determined by direct measurement of the length of the day and night; for a tablet of about this period records—

"On the sixth day of the month Nisan the day and night were equal. The day was six double-hours (*kasbu*), and the night was six double-hours."

So long as Capella was used as the indicator star, so long the year must have begun with the sun in Taurus, the Bull; but when the re-adjustment was made, and the solar tropical year connected with the equinox was substituted for the sidereal year connected with the return of the sun to a particular star, it would be seen that the association of the beginning of the year with the sun's presence in any given constellation could no longer be kept up. The necessity for an artificial division of the zodiac would be felt, and that artificial division clearly was not made until the sun at the spring equinox was unmistakably in Aries, the Ram; or about 700 B.C.

The eclipse of 1063 B.C. incidentally proves that the old method of fixing Nisan by the conjunction of the moon and Capella was then still in use; for the eclipse took place on July 31, which is called in the record "the 26th of Sivan." Sivan being the third month, its 26th day could not have fallen so late, if the year had begun with the equinox; but it would have so fallen if the Capella method were still in vogue.

There is a set of symbols repeated over and over again on Babylonian monuments, and always given a position of eminence;—it is the so-called "Triad of Stars," a crescent lying on its back and two stars near it. They are seen very distinctly at the top of the photograph of the boundary-stone from the Louvre, given on p. 318, and also immediately above the head of the Sun-god in the

WORSHIP OF THE SUN-GOD AT SIPPARA.

photograph of the tablet from Sippar, on p. 322. Their significance is now clear. Four thousand years before the Christian era, the two Twin stars, Castor and Pollux, served as indicators of the first new moon of the year, just as Capella did two thousand years later. The "triad of stars," then, is simply a picture of what men saw, year after year, in the sunset sky at the beginning of the first month, six thousand years ago. It is the earliest record of an astronomical observation that has come down to us.

How simple and easy the observation was, and how distinctly the year was marked off by it! The month was marked off by the first sight of the new thin crescent in the evening sky. The day was marked off by the return of darkness, the evening hour in which, month by month, the new moon was first observed; so that "the evening and the morning were the first day." The year was marked off by the new moon being seen in the evening with a bright pair of stars, the stars we still know as the "Twins;" and the length of the year was shown by the evening of the month, when moon and stars came together. If on the first evening, it was a year of twelve months; if on the third, one of thirteen. There was a time when these three observations constituted the whole of primitive astronomy.

In later days the original meaning of the "Triad of Stars" would seem to have been forgotten, and they were taken as representing Sin, Samas, and Istar;—the Moon, the Sun and the planet Venus. Yet now and again a hint of the part they once played in determining the length of the year is preserved. Thus, on the tablet now in the British Museum, and shown on p. 322, sculptured with a scene representing the worship of the Sun-god in the temple of Sippar, these three symbols are shown with the explanatory inscription:—

"The Moon-god, the Sun-god, and Istar, dwellers in the abyss,
Announce to the years what they are to expect;"

possibly an astrological formula, but it may well mean—"announce whether the years should expect twelve or thirteen months."

As already pointed out, this method had one drawback; it gave a sidereal year, not a tropical year, and this inconvenience must have been discovered, and Capella substituted for the Twin stars, long before the giving of the Law to Israel. The method employed by the priests of watching the progress of the ripening of the barley overcame this difficulty, and gave a year to Israel which, on the average, was a correct tropical one.

There is a detail in the history of the flood in Gen. vii. and viii. which has been taken by some as meant to indicate the length of the tropical year.

"In the six hundredth year of Noah's life, in the second month, the seventeenth day of the month, the same day were all the fountains of the great deep broken up, and the windows of heaven were opened."

"And it came to pass in the six hundredth and first year, . . . in the second month, on the seven and twentieth day of the month, was the earth dried."

The interval from the commencement of the deluge to its close was therefore twelve lunar months and ten days; *i.e.* 364 or 365 days. The beginning of the rain would, no doubt, be sharply marked; the end of the drying would be gradual, and hence the selection of a day exactly (so far as we can tell) a full tropical year from the beginning of the flood would seem to be intentional. A complete year had been consumed by the judgment.

No such total interruption of the kindly succession of the seasons shall ever occur again :—

"While the earth remaineth, seed-time and harvest, and cold and heat, and summer and winter, and day and night shall not cease."

The rain is no longer for judgment, but for blessing :—

"Thou visitest the earth, and waterest it,
Thou greatly enrichest it ;
The river of God is full of water :
Thou providest them corn, when Thou hast so prepared the earth.
Thou waterest her furrows abundantly ;
Thou settlest the ridges thereof :
Thou makest it soft with showers ;
Thou blessest the springing thereof.
Thou crownest the year with Thy goodness."

CHAPTER V

THE SABBATIC YEAR AND THE JUBILEE

THE principle of the week with its sabbath of rest was carried partially into the month, and completely into the year. The seventh month of the year was marked out pre-eminently by the threefold character of its services, though every seventh month was not distinguished. But the weekly sabbath was expressed not only in days but in years, and was one both of rest and of release.

The sabbath of years was first enjoined from Mount Sinai, in the third month after the departure from Egypt, certainly within a day or so, if not on the actual day, of the second great feast of the year, variously known to the Hebrews as the Feast of Firstfruits, or the Feast of Weeks, and to us as Pentecost, that is Whitsuntide. It is most shortly given in Exod. xxi. 2, and xxiii. 10, 11:—

"If thou buy an Hebrew servant, six years he shall serve: and in the seventh he shall go out free for nothing."

"Six years thou shalt sow thy land, and shalt gather in the fruits thereof: but the seventh year thou shalt let it rest and lie still; that the poor of thy people may eat: and what they leave the beasts of the field shall eat. In like manner thou shalt deal with thy vineyard, and with thy oliveyard."

These laws are given at greater length and with fuller explanation in the twenty-fifth chapter of the Book of Leviticus. In addition there is given a promise of blessing for the fulfilment of the laws, and, in the twenty-sixth chapter, a sign to follow on their breach.

"If ye shall say, What shall we eat the seventh year? behold, we shall not sow, nor gather in our increase: then I will command My blessing upon you in the sixth year, and it shall bring forth fruit for three years. And ye shall sow the eighth year, and eat yet of old fruit until the ninth year: until her fruits come in ye shall eat of the old store."

"Ye shall keep My sabbaths ... and if ye walk contrary unto Me ... I will scatter you among the heathen, and will draw out a sword after you: and your land shall be desolate, and your cities waste. Then shall the land enjoy her sabbaths, as long as it lieth desolate, and ye be in your enemies' land; even then shall the land rest, and enjoy her sabbaths. As long as it lieth desolate it shall rest; because it did not rest in your sabbaths, when ye dwelt upon it."

In the fifteenth chapter of the Book of Deuteronomy this sabbatic year is called a year of release. The specific injunctions here relate to loans made to a Hebrew and to a foreigner, and to the taking of a Hebrew into bondage. The laws as to loans had direct reference to the sabbath of the land, for since only Hebrews might possess the Holy Land, interest on a debt might not be exacted from a Hebrew in the sabbatic year, as the land did not then yield him wherewith he might pay. But loans to foreigners would be necessarily for commercial, not agricultural, purposes, and since commerce was not interdicted in the sabbatic year, interest on loans to foreigners might be exacted. Warning was given that the loans to a poor Hebrew should not be withheld because the sabbatic year was close at hand. The rules with respect to the Hebrew sold for debt into bondage are the same as those given in the Book of Exodus.

In Deuteronomy it was also enjoined that—

"at the end of every seven years, in the solemnity of the year of release, in the Feast of Tabernacles" (that is, in the feast of the seventh month), "when all Israel is come to appear before the Lord thy God in the place which He shall choose, thou shalt read this law before all Israel in their hearing."

We find no more mention of the sabbatic year until the reign of Zedekiah, the last king of Judah. He had made a covenant with all the people which were at Jerusalem, to proclaim liberty unto them, that every Hebrew bondservant should go free, but the princes and all the people caused their Hebrew bondservants to return and be in subjection to them. Then Jeremiah the prophet was sent to remind them of the covenant made with their fathers when they were brought out from the land of Egypt, from the house of bondmen; and in the Second Book of Chronicles it is said that the sign of the breaking of this covenant, already quoted from the Book of Leviticus, was being accomplished. The Captivity was—

"to fulfil the word of the Lord by the mouth of Jeremiah, until the land had enjoyed her sabbaths: for as long as she lay desolate she kept sabbath, to fulfil three-score and ten years."

After the exile, we find one reference to the sabbatic

year in the covenant sealed by the princes, Levites, and priests and people, in the Book of Nehemiah :—

"That we would leave the seventh year, and the exaction of every debt."

Just as the Feast of Weeks was bound to the Feast of the Passover by numbering seven sabbaths from the day of the wave-offering—" even unto the morrow after the seventh sabbath shall ye number fifty days:"—so the year of Jubilee was bound to the sabbatic year:—

"Thou shalt number seven sabbaths of years unto thee, seven times seven years; and the space of the seven sabbaths of years shall be unto thee forty and nine years. Then shalt thou cause the trumpet of the Jubile to sound on the tenth day of the seventh month, in the day of atonement shall ye make the trumpet sound throughout all your land. And ye shall hallow the fiftieth year, and proclaim liberty throughout all the land unto all the inhabitants thereof: it shall be a Jubile unto you; and ye shall return every man unto his possession and ye shall return every man unto his family."

In this year of Jubilee all land, and village houses, and the houses of the Levites were to revert to their original owners. These, in other words, could be leased only, and not bought outright, the price of the lease depending upon the number of years until the next Jubilee. A foreigner might not buy a Hebrew outright as a bondslave; he could but contract with him as a servant hired for a term; this contract might be abolished by the payment of a sum dependent on the number of years until the next year of Jubilee, and in any case the Hebrew servant and his family must go out free at the year of Jubilee. In the last chapter of the Book of Numbers we get a reference again to the year of Jubilee, and indirect allusions to it are made by Isaiah, in "the acceptable year of the Lord" when liberty should be proclaimed, and in "the year of the redeemed." In his prophecy of the restoration of Israel, Ezekiel definitely refers to "the year of liberty," when the inheritance that has been granted to a servant shall return again to the prince.

The interpretation of the sabbatic year and the year of Jubilee has greatly exercised commentators. At what season did the sabbatic year begin? was it coterminous with the ecclesiastical year; or did it differ from it by six months? Was the year of Jubilee held once in every forty-nine years or once in every fifty? did it begin at the same season as the sabbatic year? did it interrupt the reckoning of the sabbatic year, so that a new cycle commenced immediately after the year of Jubilee; or was the sabbatic year every seventh, irrespective of the year of Jubilee? did the year of Jubilee always follow immediately on a sabbatic year, or did this only happen occasionally?

The problem will be much simpler if it is borne in mind that the Law, as originally proclaimed, was eminently practical and for practical men. The period of pedantry, of hair-splitting, of slavery to mere technicalities, came very late in Jewish history.

It is clear from what has been already said in the chapter on the year, that the only calendar year in the Old Testament was the sacred one, beginning with the month

SABBATIC YEAR AND THE JUBILEE

Abib or Nisan, in the spring. At the same time the Jews, like ourselves, would occasionally refer vaguely to the beginning, or the end, or the course of the year, without meaning to set up any hard and fast connection with the authorized calendar.

Now it is perfectly clear that the sabbatic year cannot have begun with the first day of the month Abib, because the firstfruits were offered on the fifteenth of that month. That being so, the ploughing and the sowing must have taken place very considerably earlier. It is not possible to suppose that the Hebrew farmer would plough and sow his land in the last months of the previous year, knowing that he could not reap during the sabbatic year.

Similarly, it seems hardly likely that it was considered as beginning with the first of Tishri, inasmuch as the harvest festival, the Feast of the Ingathering, or Tabernacles, took place in the middle of that month. The plain and practical explanation is that, after the Feast of Tabernacles of the sixth year, the farmer would not again plough, sow, or reap his land until after the Feast of Tabernacles in the sabbatic year. The sabbatic year, in other words, was a simple agricultural year, and it did not correspond exactly with the ecclesiastical or with any calendar year.

For practical purposes the sabbatic year therefore ended with the close of the Feast of Tabernacles, when the Law was read before the whole people according to the command of Moses; and it practically began a year earlier.

The year of Jubilee appears in the directions of Lev. xxv. to have been most distinctly linked to the sabbatic year.

"The space of the seven sabbaths of years shall be unto thee forty and nine years, . . . and ye shall hallow the fiftieth year, and proclaim liberty throughout all the land unto all the inhabitants thereof: it shall be a Jubile unto you."

It would seem, therefore, that just as the week of days ran on continuously, uninterrupted by any feasts or fasts, so the week of years ran on continuously. And as the Feast of Pentecost was the 49th day from the offering of the first-fruits on the morrow of the Passover, so the Jubilee was the 49th year from the "morrow" of a sabbatic year; it followed immediately after a sabbatic year. The Jubilee was thus the 49th year from the previous Jubilee; it was the 50th from the particular sabbatic year from which the original reckoning was made.

Actually the year of Jubilee began before the sabbatic year was completed, because the trumpet of the Jubilee was to be blown upon the Day of Atonement, the 10th day of the seventh month—that is to say, whilst the sabbatic year was yet in progress. Indeed, literally speaking, this trumpet, "loud of sound," blown on the 10th day of the seventh month, *was* the Jubilee, that is to say, the sound of rejoicing, the joyful sound. A difficulty comes in here. The Israelites were commanded—

"Ye shall not sow, neither reap that which groweth of itself in it, nor gather the grapes in it of thy vine undressed. For it is the Jubile; it shall be holy unto you: ye shall eat the increase thereof out of the field."

This would appear to mean that the Jubilee extended over a whole year following a sabbatic year, so that the

SABBATIC YEAR AND THE JUBILEE

land lay fallow for two consecutive years. But this seems negatived by two considerations. It is expressly laid down in the same chapter (Lev. xxv. 22) that the Israelites were to sow in the eighth year—that is to say, in the year after a sabbatic year, and the year of Jubilee would be always a year of this character. Further, if the next sabbatic year was the seventh after the one preceding the Jubilee, then the land would be tilled for only five consecutive years, not for six, though this is expressly commanded in Lev. xxv. 3. If, on the contrary, it was tilled for six years, then the run of the sabbatic years would be interrupted.

The explanation of this difficulty may possibly be found in the fact that that which distinguished the year of Jubilee was something which did not run through the whole circuit of the seasons. The land in that year was to return to its original owners. The freehold of the land was never sold; the land was inalienable, and in the year of Jubilee it reverted. "In the year of this Jubile ye shall return every man unto his possession."

It is quite clear that it could not have been left to the caprice of the owners of property as to when this transfer took place, or as to when such Hebrews as had fallen through poverty into slavery should be liberated. If the time were made optional, grasping men would put it off till the end of the year, and sooner or later that would be the general rule. There can be no doubt that the blowing of the trumpet on the 10th day of the seventh month was the proclamation of liberty throughout all the land and to all the inhabitants thereof; and that the transfer of the land must have taken place at the same time. The slave would return to the possession of his ancestors in time to keep, as a freeman, the Feast of Tabernacles on his own land. The four days between the great day of Atonement and the Feast of Tabernacles were sufficient for this change to be carried out.

The term "Year of Jubilee" is therefore not to be taken as signifying that the events of the Jubilee were spread over twelve months, but simply, that it was the year in which the restoration of the Jubilee was accomplished. We speak of the king's "coronation year," though his coronation took place on but a single day, and the meaning that we should attach to the phrase would depend upon the particular sense in which we were using the word "year." Whilst, therefore, the Jubilee itself was strictly defined by the blowing of trumpets on the 10th day of the seventh month, it would be perfectly correct to give the title, "year of Jubilee," to any year, no matter in what season it commenced, that contained the day of that proclamation of liberty. It is also correct to say that it was the fiftieth year because it was placed at the very end of the forty-ninth year.

The difficulty still remains as to the meaning of the prohibition to sow or reap in the year of Jubilee. The command certainly reads as if the land was to lie fallow for two consecutive years; but it would seem an impracticable arrangement that the poor man returning to his inheritance should be forbidden to plough or sow until more than a twelvemonth had elapsed, and hence that he should be forbidden to reap until nearly two full years

SABBATIC YEAR AND THE JUBILEE 335

had run their course. It also, as already stated, seems directly contrary to the command to sow in the eighth year, which would also be the fiftieth. It may therefore be meant simply to emphasize the prohibition to sow and reap in the sabbatic year immediately preceding the Jubilee. The temptation would be great to a grasping man to get the most he could out of the land before parting with it for ever.

In spite of the strong array of commentators who claim that the Jubilees were to be held every fifty years as we moderns should compute it, there can be no doubt but that they followed each other at the same interval as every seventh sabbatic year; in other words, that they were held every 49 years. This is confirmed by an astronomical consideration. Forty-nine years make a convenient lunisolar cycle, reconciling the lunar month and the tropical solar year. Though not so good as the Metonic cycle of 19 years, it is quite a practical one, as the following table will show:—

3 years	=	1095·73 days	:	37 months	=	1092·63 days
8 ,,	=	2921·94 ,,	:	99 ,,	=	2923·53 ,,
11 ,,	=	4017·66 ,,	:	136 ,,	=	4016·16 ,,
19 ,,	=	6939·60 ,,	:	235 ,,	=	6939·69 ,,
49 ,,	=	17896·87 ,,	:	606 ,,	=	17895·54 ,,
60 ,,	=	21914·53 ,,	:	742 ,,	=	21911·70 ,,

The cycle of 49 years would therefore be amply good enough to guide the priestly authorities in drawing up their calendar in cases where there was some ambiguity due to the interruption of observations of the moon, and this was all that could be needed so long as the nation of Israel remained in its own land.

The cycle of 8 years is added above, since it has been stated that the Jews of Alexandria adopted this at one time from the Greeks. This was not so good as the cycle of 11 years would have been, and not to be compared with the combination of the two cycles in that of 19 years ascribed to Meton. The latter cycle was adopted by the Babylonian Jews, and forms the basis of the Jewish calendar in use to-day.

CHAPTER VI

THE CYCLES OF DANIEL

THE cycle of 49 years, marked out by the return of the Jubilee, was a useful and practical one. It supplied, in fact, all that the Hebrews, in that age, required for the purposes of their calendar. The Babylonian basic number, 60, would have given—as will be seen from the table in the last chapter—a distinctly less accurate correspondence between the month and the tropical year.

There is another way of looking at the regulations for the Jubilee, which brings out a further significant relation. On the 10th day of the first month of any year, the lamb was selected for the Passover. On the 10th day of the seventh month of any year was the great Day of Atonement. From the 10th day of the first month of the first year after a Jubilee to the next blowing of the Jubilee trumpet on the great Day of Atonement, was 600 months, that is 50 complete lunar years. And the same interval necessarily held good between the Passover of that first year and the Feast of Tabernacles of the forty-ninth year. The Passover recalled the deliverance of Israel from the bondage of Egypt; and in like manner, the release to be given to the Hebrew slave at the year of Jubilee was expressly connected with the memory of that national deliverance.

"For they are My servants, which I brought forth out of the land of Egypt: they shall not be sold as bondmen."

The day of Jubilee fell in the middle of the ecclesiastical year. From the close of the year of Jubilee—that is to say, of the ecclesiastical year in which the freeing, both of the bondmen and of the land, took place—to the next day of Jubilee was $48\frac{1}{2}$ solar years, or—as seen above—600 lunations, or 50 lunar years, so that there can be no doubt that the period was expressly designed to exhibit this cycle, a cycle which shows incidentally a very correct knowledge of the true lengths of the lunation and solar year.

This cycle was possessed by no other nation of antiquity; therefore the Hebrews borrowed it from none; and since they did not borrow the cycle, neither could they have borrowed the ritual with which that cycle was interwoven.

That the Hebrews possessed this knowledge throws some light upon an incident in the early life of the prophet Daniel.

"In the third year of the reign of Jehoiakim king of Judah came Nebuchadnezzar king of Babylon unto Jerusalem, and besieged it. . . . And the king spake unto Ashpenaz the master of his eunuchs, that he should bring certain of the children of Israel, and of the king's seed, and of the princes; children in whom was no blemish, but well favoured, and skilful in all wisdom, and cunning in knowledge, and understanding science, and such as had ability in them to stand in the king's palace, and whom they might teach the learning and the tongue of the

THE CYCLES OF DANIEL

Chaldeans. . . . Now among these were of the children of Judah, Daniel, Hananiah, Mishael, and Azariah: unto whom the prince of the eunuchs gave names: for he gave unto Daniel the name of Belteshazzar; and to Hananiah, of Shadrach; and to Mishael, of Meshach; and to Azariah, of Abed-nego. . . . As for these four children, God gave them knowledge and skill in all learning and wisdom: and Daniel had understanding in all visions and dreams. Now at the end of the days that the king had said he should bring them in, then the prince of the eunuchs brought them in before Nebuchadnezzar. And the king communed with them; and among them all was found none like Daniel, Hananiah, Mishael, and Azariah: therefore stood they before the king. And in all matters of wisdom and understanding, that the king inquired of them, he found them ten times better than all the magicians and astrologers that were in all his realm."

The Hebrew children that king Nebuchadnezzar desired to be brought were to be already possessed of knowledge; they were to be further instructed in the learning and tongue of the Chaldeans. But when the four Hebrew children were brought before the king, and he communed with them, he found them wiser than his own wise men.

No account is given of the questions asked by the king, or of the answers made by the four young Hebrews; so it is merely a conjecture that possibly some question bearing on the calendar may have come up. But if it did, then certainly the information within the grasp of the Hebrews could not have failed to impress the king.

We know how highly the Greeks esteemed the discovery by Meton, in the 86th Olympiad, of that relation between the movements of the sun and moon, which gives the cycle of nineteen years, and similar knowledge would certainly have given king Nebuchadnezzar a high opinion of the young captives.

But there is evidence, from certain numbers in the book which bears his name, that Daniel was acquainted with luni-solar cycles which quite transcended that of the Jubilees in preciseness, and indicate a knowledge such as was certainly not to be found in any other ancient nation. The numbers themselves are used in a prophetic context, so that the meaning of the whole is veiled, but astronomical knowledge underlying the use of these numbers is unmistakably there.

One of these numbers is found in the eighth chapter.

"How long shall be the vision concerning the daily sacrifice, and the transgression of desolation, to give both the sanctuary and the host to be trodden under foot? And he said unto me, Unto two thousand and three hundred days; then shall the sanctuary be cleansed."

The twelfth chapter gives the other number, but in a more veiled form :—

"And I heard the man clothed in linen, which was upon the waters of the river, when he held up his right hand and his left hand unto heaven, and sware by Him that liveth for ever that it shall be for a time, times, and an half; and when he shall have accomplished to scatter the power of the holy people, all these things shall be finished."

The numerical significance of the "time, times and an half," or, as it is expressed in the seventh chapter of Daniel, "until a time, and times, and the dividing of time," is plainly shown by the corresponding expressions in the Apocalypse, where "a time and times and half a time"

THE CYCLES OF DANIEL 341

would appear to be given elsewhere both as "forty and two months" and "a thousand, two hundred and threescore days." Forty-two conventional months—that is of 30 days each—make up 1260 days, whilst $3\frac{1}{2}$ conventional years of 360 days—that is twelve months of 30 days each—make up the same period. The word "times" is expressly used as equivalent to years in the eleventh chapter of Daniel, where it is said that the king of the north "shall come on at the end of the times, even of years, with a great army and with much substance." Then, again in the vision which Nebuchadnezzar had previous to his madness, he heard the watcher and the holy one cry concerning him :—

"Let his heart be changed from man's, and let a beast's heart be given unto him; and let seven times pass over him."

It has been generally understood that the "seven times" in this latter case meant "seven years." The "time, times and an half" are obviously meant as the half of "seven times."

The two numbers, 2,300 and 1,260, whatever be their significance in their particular context in these prophecies, have an unmistakable astronomical bearing, as the following table will show :—

```
    2,300 solar years    = 840,057 days, 1 hour.
   28,447 lunar months  = 840,056   „  16 hours.
           difference   =              9    „
    1,260 solar years    = 460,205   „   4    „
   15,584 lunar months  = 460,204   „  17    „
           difference   =             11    „
```

If the one number 1,260 stood alone, the fact that it was so close a lunar cycle might easily be ascribed to a mere coincidence. Seven is a sacred number, and the days in the year may be conventionally represented as 360. Half the product of the two might, perhaps, seem to be a natural number to adopt for symbolic purposes. But the number 2,300 stands in quite a different category. It is not suggested by any combination of sacred numbers, and is not veiled under any mystic expression; the number is given as it stands—2,300. But 2,300 solar years is an exact number, not only of lunations, but also of "anomalistic" months. The "anomalistic month" is the time occupied by the moon in travelling from its perigee, that is its point of nearest approach to the earth, round to its perigee again. For the moon's orbit round the earth is not circular, but decidedly elliptical; the moon being 31,000 miles nearer to us at perigee than it is at apogee, its point of greatest distance. But it moves more rapidly when near perigee than when near apogee, so that its motion differs considerably from perfect uniformity.

But the period in which the moon travels from her perigee round to perigee again is 27 days, 13 hours, 18 minutes, 37 seconds, and there are in 2,300 solar years almost exactly 30,487 such periods or anomalistic months, which amount to 840,057 days, 2 hours.

If we take the mean of these three periods, that is to say 840,057 days, as being the cycle, it brings into harmony the day, the anomalistic month, the ordinary month, and the solar year. It is from this point of view the most perfect cycle known.

THE CYCLES OF DANIEL

Dr. H. Grattan Guinness [1] has shown what a beautifully simple and accurate calendar could have been constructed on the basis of this period of 2,300 years; thus:—

2,300 solar years contain 28,447 synodic months, of which 847 are intercalary, or epact months. 2,300 years are 840,057 days:—

$$
\begin{array}{r}
\text{Days.} \\
27{,}600 \begin{cases} 13{,}800 \text{ non-intercalary mths. of } 29 \text{ d. each} = 400{,}200 \\ 13{,}800 \quad ,, \qquad ,, \qquad ,, \quad ,, \; 30 \quad ,, \quad = 414{,}000 \end{cases} \\
847 \begin{cases} 423 \text{ intercalary months of } 30 \text{ days each} = 12{,}690 \\ 424 \quad ,, \qquad ,, \qquad ,, \; 31 \quad ,, \quad ,, \quad = 13{,}144 \end{cases} \\
23 \text{ days additional for the 23 centuries} = 23 \\
\hline
840{,}057
\end{array}
$$

The Jewish calendar on this system would have consisted of ordinary months, alternately 29 and 30 days in length. The intercalary months would have contained alternately 30 or 31 days, and once in every century one of the ordinary months would have had an additional day. Or, what would come to very much the same thing, this extra day might have been added at every alternate Jubilee.

By combining these two numbers of Daniel some cycles of extreme astronomical interest have been derived by De Cheseaux, a Swiss astronomer of the eighteenth century, and by Dr. H. Grattan Guinness, and Dr. W. Bell Dawson in our own times. Thus, the difference between 2,300 and 1,260 is 1,040, and 1,040 years give an extremely exact correspondence between the solar year and the month, whilst the mean of the two numbers gives us 1,780, and

---

[1] *Creation centred in Christ*, p. 344.

1,780 lunar years is 1,727 solar years with extreme precision. But since these are not given directly in the Book of Daniel, and are only inferential from his numbers, there seems no need to comment upon them here.

It is fair, however, to conclude that Daniel was aware of the Metonic cycle. The 2300-year cycle gives evidence of a more accurate knowledge of the respective lengths of month and year than is involved in the cycle of 19 years. And the latter is a cycle which a Jew would be naturally led to detect, as the number of intercalary months contained in it is seven, the Hebrew sacred number.

The Book of Daniel, therefore, itself proves to us that king Nebuchadnezzar was perfectly justified in the high estimate which he formed of the attainments of the four Hebrew children. Certainly one of them, Daniel, was a better instructed mathematician and astronomer than any Chaldean who had ever been brought into his presence.

We have the right to make this assertion, for now we have an immense number of Babylonian records at our command; and can form a fairly accurate estimate as to the state there of astronomical and mathematical science at different epochs. A kind of "quasi-patriotism" has induced some Assyriologists to confuse in their accounts of Babylonian attainments the work of times close to the Christian era with that of many centuries, if not of several millenniums earlier; and the times of Sargon of Agadé, whose reputed date is 3800 B.C., have seemed to be credited with the astronomical work done in Babylon in the first and second centuries before our era. This is much as if

THE CYCLES OF DANIEL

we should credit our predecessors who lived in this island at the time of Abraham with the scientific attainments of the present day.

The earlier astronomical achievements at Babylon were not, in any real sense, astronomical at all. They were simply the compilation of lists of crude astrological omens, of the most foolish and unreasoning kind. Late in Babylonian history there were observations of a high scientific order; real observations of the positions of moon and planets, made with great system and regularity. But these were made after Greek astronomy had attained a high level, and Babylon had come under Greek rule.

Whether this development of genuine astronomical observation was of native origin, or was derived from their Greek masters, is not clear. If it was native, then certainly the Babylonians were not able to use and interpret the observations which they made nearly so well as were Greek astronomers, such as Eudoxus, Thales, Pythagoras, Hipparchus and many others.

But it must not be supposed that, though their astronomical achievements have been grossly, even ludicrously, exaggerated by some popular writers, the Babylonians contributed nothing of value to the progress of the science. We may infer from such a tablet as that already quoted on page 320, when the equinox was observed on the 6th day of Nisan, since there were 6 *kasbu* of day and 6 *kasbu* of night, that some mechanical time-measurer was in use. Indeed, the record on one tablet has been interpreted as noting that the astronomer's clock or clepsydra had stopped. If this be so, then we owe to

Babylon the invention of clocks of some description, and from an astronomical point of view, this is of the greatest importance.

Tradition also points to the Chaldeans as the discoverers of the *Saros*, the cycle of 18 years, 10 or 11 days, after which eclipses of the sun or moon recur. The fact that very careful watch was kept every month at the times of the new and of the full moon, at many different stations, to note whether an eclipse would take place, would naturally bring about the discovery of the period, sooner or later.

The achievements of a nation will be in accordance with its temperament and opportunities, and it is evident from the records which they have left us that the Babylonians, though very superstitious, were a methodical, practical, prosaic people, and a people of that order, if they are numerous, and under strong rule, will go far and do much. The discovery of the *Saros* was such as was within their power, and was certainly no small achievement. But it is to the Greeks, not to the Babylonians, that we trace the beginnings of mathematics and planetary theory.

We look in vain amongst such Babylonian poetry as we possess for the traces of a Homer, a Pindar, a Sophocles, or even of a poet fit to enter into competition with those of the second rank in the literature of Greece; while it must remain one of the literary mysteries of our time that any one should deem the poetry of the books of Isaiah and Job dependent on Babylonian inspiration.

There were two great hindrances under which the Babylonian man of science laboured: he was an idolater, and he was an astrologer. It is not possible for us in our freedom to fully realize how oppressive was the slavery of mind, as well as spirit, which was consequent upon this twofold superstition. The Greek was freer, insomuch that he did not worship the planets, and did not become a planetary astrologer until after he had learnt that superstition from Chaldea; in learning it he put an end to his scientific progress.

But the Hebrew, if he was faithful to the Law that had been given to him, was free in mind as well as in spirit. He could fearlessly inquire into any and all the objects of nature, for these were but things—the work of God's Hands, whereas he, made in the image of God, having the right of intercourse with God, was the superior, the ruler of everything he could see.

His religious attitude therefore gave him a great superiority for scientific advancement. Yet there was one phase of that attitude which, whilst it preserved him from erroneous conceptions, tended to check that spirit of curiosity which has led to so much of the scientific progress of modern times. "What?" "How?" and "Why?" are the three questions which man is always asking of nature, and to the Hebrew the answer to the second and third was obvious :—It is the power of God : It is the will of God. He did not need to invent for himself the crass absurdities of the cosmogonies of the heathen; but neither was he induced to go behind the appearances of things; the sufficient cause and explanation of all was God.

But of the appearances he was very observant, as I trust has become clear in the course of this imperfect review of the traces of one particular science as noticed in Holy Scripture.

If he was faithful to the Law which had been given him, the Hebrew was free in character as well as in mind. His spirit was not that of a bondman, and Nebuchadnezzar certainly never met anything more noble, anything more free, than the spirit of the men who answered him in the very view of the burning fiery furnace :—

"O Nebuchadnezzar, we are not careful to answer thee in this matter. If it be so, our God whom we serve is able to deliver us from the burning fiery furnace, and He will deliver us out of thine hand, O king. BUT IF NOT, be it known unto thee, O king, that we will not serve thy gods, nor worship the golden image which thou hast set up."

"SUN, STAND THOU STILL UPON GIBEON, AND THOU MOON IN THE VALLEY OF AJALON."

BOOK IV

THREE ASTRONOMICAL MARVELS

CHAPTER I

JOSHUA'S "LONG DAY"[1]

1.—Method of Studying the Record

THERE are three incidents recorded in Holy Scripture which may fairly, if with no great exactness, be termed astronomical miracles;—the "long day" on the occasion of Joshua's victory at Beth-horon; the turning back of the shadow on the dial of Ahaz, as a sign of king Hezekiah's recovery from sickness; and the star which guided the wise men from the east to the birthplace of the Holy Child at Bethlehem.

As astronomy has some bearing on each of these three remarkable events, it will be of some interest to examine each of them from the point of view of our present astronomical knowledge. It does not follow that this will throw any new light upon the narratives, for we must always bear in mind that the Scriptures were not intended to teach us the physical

[1] Revised and reprinted from the *Sunday at Home* for February and March, 1904.

352   THE ASTRONOMY OF THE BIBLE

sciences; consequently we may find that the very details have been omitted which an astronomer, if he were writing an account of an astronomical observation, would be careful to preserve. And we must further remember that we have not the slightest reason to suppose that the sacred historians received any supernatural instruction in scientific matters. Their knowledge of astronomy therefore was that which they had themselves acquired from education and research, and nothing more. In other words, the astronomy of the narrative must be read strictly in the terms of the scientific advancement of the writers.

But there is another thing that has also to be remembered. The narrative which we have before us, being the only one that we have, must be accepted exactly as it stands. That is the foundation of our inquiry; we have no right to first cut it about at our will, to omit this, to alter that, to find traces of two, three, or more original documents, and so to split up the narrative as it stands into a number of imperfect fragments, which by their very imperfection may seem to be more or less in conflict.

The scientific attitude with regard to the record of an observation cannot be too clearly defined. If that record be the only one, then we may accept it, we may reject it, we may be obliged to say, "We do not understand it," or "It is imperfect, and we can make no use of it," but we must not alter it. A moment's reflection will show that a man who would permit himself to tamper with the sole evidence upon which he purports to work, no matter how profoundly convinced he may be that his proposed corrections are sound, is one who does not understand the spirit of science, and *is* not going the way to arrive at scientific truth.

There is no need then to inquire as to whether the tenth chapter of the Book of Joshua comes from two or more sources; we take the narrative as it stands. And it is one which has, for the astronomer, an interest quite irrespective of any interpretation which he may place upon the account of the miracle which forms its central incident. For Joshua's exclamation:—

"Sun, stand thou still upon Gibeon;
And thou, Moon, in the valley of Ajalon,"

implies that, at the moment of his speaking, the two heavenly bodies appeared to him to be, the one upon or over Gibeon, the other over the valley of Ajalon. We have therefore, in effect, a definite astronomical observation; interesting in itself, as being one of the oldest that has been preserved to us; doubly interesting in the conclusions that we are able to deduce from it.

The idea which has been most generally formed of the meaning of Joshua's command, is, that he saw Gibeon in the distance on the horizon in one direction with the sun low down in the sky immediately above it, and the valley of Ajalon in the distance, on the horizon in another direction, with the moon low down in the sky above it.

It would be quite natural to associate the sun and moon with distant objects if they were only some five or six degrees high; it would be rather straining the point to do so if they were more than ten degrees high; and if they were fifteen or more degrees high, it would be quite impossible.

They could not be both in the same quarter of the sky; both rising or both setting. For this would mean that the moon was not only very near the sun in the sky, but was very near to conjunction—in other words, to new moon. She could, therefore, have only shown a slender thread of light, and it is perfectly certain that Joshua, facing the sun in such a country as southern Palestine could not possibly have perceived the thin pale arch of light, which would have been all that the moon could then have presented to him. Therefore the one must have been rising and the other setting, and Joshua must have been standing between Gibeon and the valley of Ajalon, so that the two places were nearly in opposite directions from him. The moon must have been in the west and the sun in the east, for the valley of Ajalon is west of Gibeon. That is to say, it cannot have been more than an hour after sunrise, and it cannot have been more than an hour before moonset. Adopting therefore the usual explanation of Joshua's words, we see at once that the common idea of the reason for Joshua's command to the sun, namely, that the day was nearly over, and that he desired the daylight to be prolonged, is quite mistaken. If the sun was low down in the sky, he would have had practically the whole of the day still before him.

2.—BEFORE THE BATTLE

Before attempting to examine further into the nature of the miracle, it will be well to summarize once again the familiar history of the early days of the Hebrew invasion of Canaan. We are told that the passage of the Jordan took place on the tenth day of the first month; and that the Feast of the Passover was held on the fourteenth day of that month. These are the only two positive dates given us. The week of the Pascal celebrations would have occupied the time until the moon's last quarter. Then preparations were made for the siege of Jericho, and another week passed in the daily processions round the city before the moment came for its destruction, which must have been very nearly at the beginning of the second month of the year. Jericho having been destroyed, Joshua next ordered a reconnaissance of Aï, a small fortified town, some twenty miles distant, and some 3400 feet above the Israelite camp at Gilgal, and commanding the upper end of the valley of Achor, the chief ravine leading up from the valley of the Jordan. The reconnaissance was followed by an attack on the town, which resulted in defeat. From the dejection into which this reverse had thrown him Joshua was roused by the information that the command to devote the spoil of Jericho to utter destruction had been disobeyed. A searching investigation was held; it was found that Achan, one of the Israelite soldiers, had seized for himself a royal robe and an ingot of gold; he was tried, condemned and executed, and the army of Israel was absolved from his guilt. A second attack was made upon Aï; the town was taken; and the road was made clear for Israel to march into the heart of the country, in order to hold the great religious ceremony of the reading of the law upon the mountains of Ebal and Gerizim, which had been commanded them long before. No note is given of the date when this ceremony took place, but bearing in mind that the second month of the year must have begun at the time of the first reconnaissance of Aï, and that the original giving of the Law upon Mount Sinai had taken place upon the third day of the third month, it seems most likely that that anniversary would be chosen for a solemnity which was intended to recall the original promulgation in the most effective manner. If this were so, it would account for the circumstance, which would otherwise have seemed so strange, that Joshua should have attacked two cities only, Jericho and Aï, and then for a time have held his hand. It was the necessity of keeping the great national anniversary on the proper day which compelled him to desist from his military operations after Aï was taken.

We are not told how long the religious celebrations at Shechem lasted, but in any case the Israelites can hardly have been back in their camp at Gilgal before the third moon of the year was at the full. But after their return, events must have succeeded each other with great rapidity. The Amorites must have regarded the pilgrimage of Israel to Shechem as an unhoped-for respite, and they took advantage of it to organize a great confederacy. Whilst this confederacy was being formed, the rulers of a small state of "Hivites"—by which we must understand a community differing either in race or habits from the generality of their Amorite neighbours—had been much exercised by the course of events. They had indeed reason to be. Aï, the last

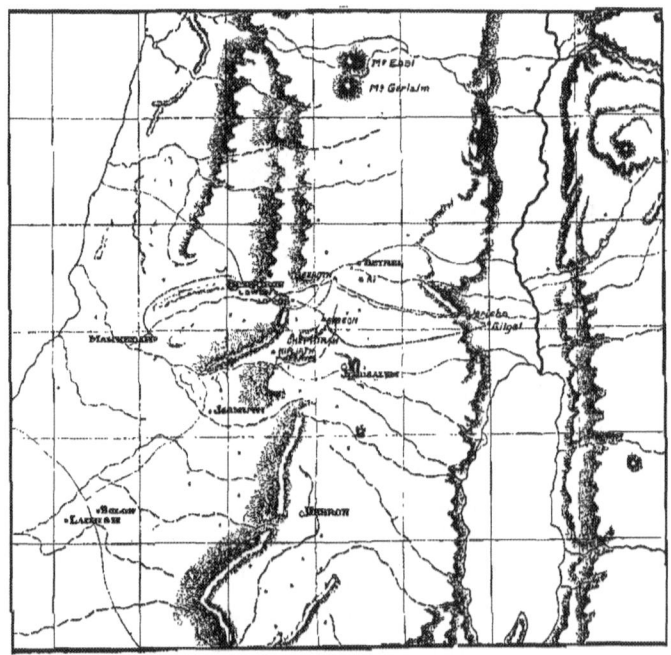

MAP OF SOUTHERN PALESTINE.
Amorite Cities, thus: HEBRON. Hivite Cities: BEEROTH.
Places taken by the Israelites: Jericho.
Conjectural line of march of Joshua: ....

conquest of Israel, was less than four miles, as the crow flies, from Bireh, which is usually identified with Beeroth, one of the four cities of the Hivite State; and the Beerothites had, without doubt, watched the cloud of smoke go up from the burning town when it was sacked; and the mound which now covered what had been so recently their neighbour city, was visible almost from their gates. That was an object-lesson which required no enforcement. The Hivites, sure that otherwise their turn would come next, resolved to make peace with Israel before they were attacked.

To do this they had to deceive the Israelites into believing that they were inhabitants of some land far from Canaan, and this they must do, not only before Joshua actually attacked them, but before he sent out another scouting party. For Beeroth would inevitably have been the very first town which it would have approached, and once Joshua's spies had surveyed it, all chance of the Hivites successfully imposing upon him would have vanished.

But they were exposed to another danger, if possible more urgent still. The headquarters of the newly formed Amorite league was at Jerusalem, on the same plateau as Gibeon, the Hivite capital, and distant from it less than six miles. A single spy, a single traitor, during the anxious time that their defection was being planned, and Adoni-zedec, the king of Jerusalem, would have heard of it in less than a couple of hours; and the Gibeonites would have been overwhelmed before Joshua had any inkling that they were anxious to treat with him. Whoever was dilatory, whoever was slow, the Gibeonites dared not be. It can, therefore, have been, at most, only a matter of hours after Joshua's return to Gilgal, before their wily embassy set forth.

But their defection had an instant result. Adoni-zedec recognized in a moment the urgency of the situation. With Joshua in possession of Gibeon and its dependencies, the Israelites would be firmly established on the plateau at his very gates, and the states of southern Palestine would be cut off from their brethren in the north.

Adoni-zedec lost no time; he sought and obtained the aid of four neighbouring kings and marched upon Gibeon. The Gibeonites sent at once the most urgent message to acquaint Joshua with their danger, and Joshua as promptly replied. He made a forced march with picked troops all that night up from Gilgal, and next day he was at their gates.

Counterblow had followed blow, swift as the clash of rapiers in a duel of fencers. All three of the parties concerned—Hivite, Amorite and Israelite—had moved with the utmost rapidity. And no wonder; the stake for which they were playing was very existence, and the forfeit, which would be exacted on failure, was extinction.

3.—Day, Hour, and Place of the Miracle

The foregoing considerations enable us somewhat to narrow down the time of the year at which Joshua's miracle can have taken place, and from an astronomical point of view this is very important. The Israelites had entered the land of Canaan on the 10th day of the first month, that is to say, very shortly after the spring equinox—March 21 of our present calendar. Seven weeks after that equinox—May 11—the sun attains a declination of

18° north. From this time its declination increases day by day until the summer solstice, when, in Joshua's time, it was nearly 24° north. After that it slowly diminishes, and on August 4 it is 18° again. For twelve weeks, therefore—very nearly a quarter of the entire year—the sun's northern declination is never less than 18°. The date of the battle must have fallen somewhere within this period. It cannot have fallen earlier; the events recorded could not possibly have all been included in the seven weeks following the equinox. Nor, in view of the promptitude with which all the contending parties acted, and were bound to act, can we postpone the battle to a later date than the end of this midsummer period.

We thus know, roughly speaking, what was the declination of the sun—that is to say, its distance from the equator of the heavens—at the time of the battle; it was not less than 18° north of the equator, it could not have been more than 24°.

But, if we adopt the idea most generally formed of the meaning of Joshua's command, namely, that he saw the sun low down over Gibeon in one direction, and the moon low down over the valley of Ajalon in another, we can judge of the apparent bearing of those two heavenly bodies from an examination of the map. And since, if we may judge from the map of the Palestine Exploration Fund, the valley of Ajalon lies about 17° north of west from Gibeon, and runs nearly in that direction from it, the moon must, to Joshua, have seemed about 17° north of west, and the sun 17° south of east.

But for any date within the three summer months, the

JOSHUA'S 'LONG DAY'

sun in the latitude of Gibeon, when it bears 17° south of east, must be at least 56° high. At this height it would seem overhead, and would not give the slightest idea of association with any distant terrestrial object. Not until some weeks after the autumnal equinox could the sun be seen low down on the horizon in the direction 17° south of east, and at the same time the moon be as much as 17° north of the west point. And, as this would mean that the different combatants had remained so close to each other, some four or five months without moving, it is clearly inadmissible. We are forced therefore to the unexpected conclusion that *it is practically impossible that Joshua could have been in any place from whence he could have seen, at one and the same moment, the sun low down in the sky over Gibeon, and the moon over the valley of Ajalon.*

Is the narrative in error, then? Or have we been reading into it our own erroneous impression? Is there any other sense in which a man would naturally speak of a celestial body as being " over " some locality on the earth, except when both were together on his horizon?

Most certainly. There is another position which the sun can hold in which it may naturally be said to be " over," or " upon " a given place; far more naturally and accurately than when it chances to lie in the same direction as some object on the horizon. We have no experience of that position in these northern latitudes, and hence perhaps our commentators have, as a rule, not taken it into account. But those who, in tropical or sub-tropical countries, have been in the open at high noon, when a man's foot can almost cover his shadow, will recognize how definite, how significant such a position is. In southern Palestine, during the three summer months, the sun is always so near the zenith at noon that it could never occur to any one to speak of it as anything but "overhead."

And the prose narrative expressly tells us that this was the case. It is intimated that when Joshua spoke it was noon, by the expression that the sun "hasted not to go down about a whole day," implying that the change in the rate in its apparent motion occurred only in the afternoon, and that it had reached its culmination. Further, as not a few commentators have pointed out, the expression,—"the sun stood still in the midst of heaven,"—is literally "in the bisection of heaven"; a phrase applicable indeed to any position on the meridian, but especially appropriate to the meridian close to the zenith.

This, then, is what Joshua meant by his command to the sun. Its glowing orb blazed almost in the centre of the whole celestial vault—"in the midst of heaven"—and poured down its vertical rays straight on his head. It stood over him—it stood over the place where he was—Gibeon.

We have, therefore, been able to find that the narrative gives us, by implication, two very important particulars, the place where Joshua was, and the time of the day. He was at Gibeon, and it was high noon.

The expression, "Thou, Moon, in the valley of Ajalon," has now a very definite signification. As we have already seen, the valley of Ajalon bears 17° north of west from Gibeon, according to the map of the Palestine Exploration

JOSHUA'S 'LONG DAY'

Fund, so that this is the azimuth which the moon had at the given moment. In other words, it was almost exactly midway between the two "points of the compass," W.b.N. and W.N.W. It was also in its "last quarter" or nearly so; that is, it was half-full, and waning. With the sun on the meridian it could not have been much more than half-full, for in that case it would have already set; nor

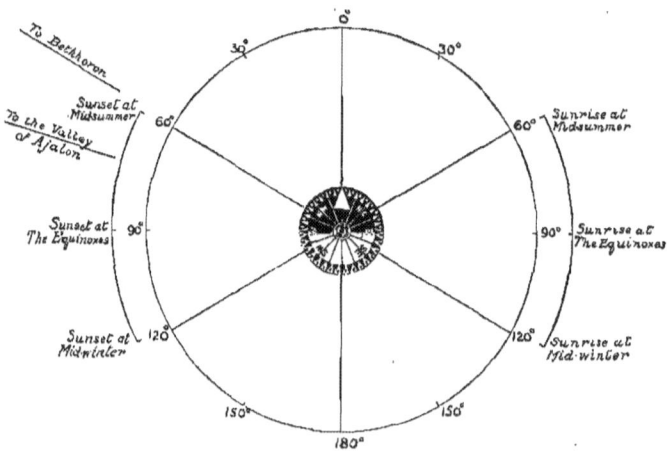

BEARINGS OF THE RISING AND SETTING POINTS OF THE SUN FROM GIBEON.

much less than half-full, or it would have been too faint to be seen in full daylight. It was therefore almost exactly half-full, and the day was probably the 21st day of the month in the Jewish reckoning.

But the moon cannot be as far as 17° north of west in latitude 31° 51′ N. on the 21st day of the month earlier than the fourth month of the Jewish year, or later than the eighth month. Now the 21st day of the fourth month is about seven weeks after the 3rd day of the third month; the 21st day of the fifth month is eleven weeks. Remembering how close Gilgal, Gibeon and Jerusalem were to each other, and how important was the need for promptitude to Israelite and Amorite alike, it can scarcely be disputed that eleven weeks is an inadmissible length of time to interpose between the reading of the Law and the battle; and that seven weeks is the utmost that can be allowed.

The battle took place, then, on or about the 21st day of the fourth month. But it could only have done so if that particular year began late. If the year had begun earlier than April 1st of our present calendar, the moon could not have been so far north on the day named. For the Jewish calendar is a natural one and regulated both by the sun and the moon. It begins with the new moon, and it also begins as nearly as possible with the spring equinox. But as twelve natural months fall short of a solar year by eleven days, a thirteenth month has to be intercalated from time to time; in every nineteen years, seven are years having an extra month. Now the 21st day of the fourth month must have fallen on or about July 22 according to our present reckoning, in order that the moon might have sufficient northing, and that involves a year beginning after April 1; so that the year of the battle of Beth-horon must have been an ordinary year, one of twelve months, but must have followed a year of thirteen months.

Summarizing all the conclusions at which we have now

arrived, Joshua's observation was made at Gibeon itself, almost precisely at the moment of noon, on or about the 21st day of the fourth month, which day fell late in July according to our present reckoning; probably on or about the 22nd. The sun's declination must have been about 20° north; probably, if anything, a little more. The sun rose therefore almost exactly at five in the morning, and set almost exactly at seven in the evening, the day being just fourteen hours long. The moon had not yet passed her third quarter, but was very near it; that is to say, she was about half full. Her declination did not differ greatly from 16° north; she was probably about 5° above the horizon, and was due to set in about half an hour. She had risen soon after eleven o'clock the previous evening, and had lighted the Israelites during more than half of their night march up from Gilgal.

4.—Joshua's Strategy

These conclusions, as to the place and time of day, entirely sweep away the impression, so often formed, that Joshua's victory was practically in the nature of a night surprise. Had it been so, and had the Amorites been put to flight at daybreak, there would have seemed no conceivable reason why, with fourteen hours of daylight before him, Joshua should have been filled with anxiety for the day to have been prolonged. Nor is it possible to conceive that he would still have been at Gibeon at noon, seven hours after he had made his victorious attack upon his enemy.

The fact is that, in all probability, Joshua had no wish

to make a night surprise. His attitude was like that of Nelson before the battle of Trafalgar; he had not the slightest doubt but that he would gain the victory, but he was most anxious that it should be a complete one. The great difficulty in the campaign which lay before him was the number of fortified places in the hands of the enemy, and the costliness, both in time and lives, of all siege operations at that epoch. His enemies having taken the field gave him the prospect of overcoming this difficulty, if, now that they were in the open, he could succeed in annihilating them there; to have simply scattered them would have brought him but little advantage. That this was the point to which he gave chief attention is apparent from one most significant circumstance in the history; the Amorites fled by the road to Beth-horon.

There have been several battles of Beth-horon since the days of Joshua, and the defeated army has, on more than one occasion, fled by the route now taken by the Amorites. Two of these are recorded by Josephus; the one in which Judas Maccabæus defeated and slew Nicanor, and the other when Cestius Gallus retreated from Jerusalem. It is probable that Beth-horon was also the scene of one, if not two, battles with the Philistines, at the commencement of David's reign. In all these cases the defeated foe fled by this road because it had been their line of advance, and was their shortest way back to safety.

But the conditions were entirely reversed in the case of Joshua's battle. The Amorites fled *away from* their cities. Jerusalem, the capital of Adoni-zedec and the chief city of the confederation, lay in precisely the opposite direction.

The other cities of their league lay beyond Jerusalem, further still to the south.

A reference to the map shows that Gilgal, the headquarters of the army of Israel, was on the plain of Jericho, close to the banks of the Jordan, at the bottom of that extraordinary ravine through which the river runs. Due west, at a distance of about sixteen or seventeen miles as the crow flies, but three thousand four hundred feet above the level of the Jordan, rises the Ridge of the Watershed, the backbone of the structure of Palestine. On this ridge are the cities of Jerusalem and Gibeon, and on it, leading down to the Maritime Plain, runs in a north-westerly direction, the road through the two Beth-horons.

The two Beth-horons are one and a half miles apart, with a descent of 700 feet from the Upper to the Lower.

The flight of the Amorites towards Beth-horon proves, beyond a doubt, that Joshua had possessed himself of the road from Gibeon to Jerusalem. It is equally clear that this could not have been done by accident, but that it must have been the deliberate purpose of his generalship. Jerusalem was a city so strong that it was not until the reign of David that the Israelites obtained possession of the whole of it, and to take it was evidently a matter beyond Joshua's ability. But to have defeated the Amorites at Gibeon, and to have left open to them the way to Jerusalem—less than six miles distant—would have been a perfectly futile proceeding. We may be sure, therefore, that from the moment when he learned that Adoni-zedek was besieging Gibeon, Joshua's first aim was to cut off the Amorite king from his capital.

368 THE ASTRONOMY OF THE BIBLE

The fact that the Amorites fled, not towards their cities but away from them, shows clearly that Joshua had specially manœuvred so as to cut them off from Jerusalem. How he did it, we are not told, and any explanation offered must necessarily be merely of the nature of surmise. Yet a considerable amount of probability may attach to it. The geographical conditions are perfectly well known, and we can, to some degree, infer the course which the battle must have taken from these, just as we could infer the main lines of the strategy employed by the Germans in their war with the French in 1870, simply by noting the places where the successive battles occurred. The positions of the battlefields of Mars-la-Tour, Gravelotte, and Sedan would show clearly that the object of the Germans had been, first, to shut Bazaine up in Metz, and then to hinder MacMahon from coming to his relief. So in the present case, the fact that the Amorites fled by the way of the two Beth-horons, shows, first, that Joshua had completely cut them off from the road to Jerusalem, and next, that somehow or other when they took flight they were a long way to the north of him. Had they not been so, they could not have had any long start in their flight, and the hailstorm which occasioned them such heavy loss would have injured the Israelites almost as much.

How can these two circumstances be accounted for? I think we can make a very plausible guess at the details of Joshua's strategy from noting what he is recorded to have done in the case of Aï. On that occasion, as on this, he had felt his inability to deal with an enemy behind fortifications. His tactics therefore had consisted in making a feigned attack, followed by a feigned retreat, by which he drew his enemies completely away from their base, which he then seized by means of a detachment which he had previously placed in ambush near. Then, when the men of Aï were hopelessly cut off from their city, he brought all his forces together, surrounded his enemies in the open, and destroyed them.

It was a far more difficult task which lay before him at Gibeon, but we may suppose that he still acted on the same general principles. There were two points on the ridge of the watershed which, for very different reasons, it was important that he should seize. The one was Beeroth, one of the cities of the Hivites, his allies, close to his latest victory of Aï, and commanding the highest point on the ridge of the watershed. It is distant from Jerusalem some ten miles—a day's journey. Tradition therefore gives it as the place where the Virgin Mary and St. Joseph turned back sorrowing, seeking Jesus. For " they, supposing Him to have been in the company, went a day's journey," and Beeroth still forms the first halting-place for pilgrims from the north on their return journey.

Beeroth also was the city of the two sons of Rimmon who murdered Ishbosheth, the son of Saul. When it is remembered how Saul had attempted to extirpate the Gibeonites, and how bitter a blood feud the latter entertained against his house in consequence, it becomes very significant that the murderers of his son were men of this Gibeonite town.

Beeroth also commanded the exit from the principal ravine by which Joshua could march upwards to the ridge

—the valley of Achor. The Israelites marching by this route would have the great advantage that Beeroth, in the possession of their allies, the Gibeonites, would act as a cover to them whilst in the ravines, and give them security whilst taking up a position on the plateau.

But Beeroth had one fatal disadvantage as a sole line of advance. From Beeroth Joshua would come down to Gibeon from the north, and the Amorites, if defeated, would have a line of retreat, clear and easy, to Jerusalem. It was absolutely essential that somewhere or other he should cut the Jerusalem road.

This would be a matter of great difficulty and danger, as, if his advance were detected whilst he was still in the ravines, he would have been taken at almost hopeless disadvantage. The fearful losses which the Israelites sustained in the intertribal war with Benjamin near this very place, show what Joshua might reasonably have expected had he tried to make his sole advance on the ridge near Jerusalem.

Is it not probable that he would have endeavoured, under these circumstances, to entice the Amorites as far away to the north as possible before he ventured to bring his main force out on the ridge? If so, we may imagine that he first sent a strong force by the valley of Achor to Beeroth; that they were instructed there to take up a strong position, and when firmly established, to challenge the Amorites to attack them. Then, when the Israelite general in command at Beeroth perceived that he had before him practically the whole Amorite force—for it would seem clear that the five kings themselves, together with the greater part of their army, were thus drawn away—he would signal to Joshua that the time had come for his advance. Just as Joshua himself had signalled with his spear at the taking of Aï, so the firing of a beacon placed on the summit of the ridge would suffice for the purpose. Joshua would then lead up the main body, seize the Jerusalem road, and press on to Gibeon at the utmost speed. If this were so, the small detachment of Amorites left to continue the blockade was speedily crushed, but perhaps was aware of Joshua's approach soon enough to send swift runners urging the five kings to return. The news would brook no delay; the kings would turn south immediately; but for all their haste they never reached Gibeon. They probably had but advanced as far as the ridge leading to Beth-horon, when they perceived that not only had Joshua relieved Gibeon and destroyed the force which they had left before it, but that his line, stretched out far to the right and left, already cut them off, not merely from the road to Jerusalem and Hebron, but also from the valley of Ajalon, a shorter road to the Maritime Plain than the one they actually took. East there was no escape; north was the Israelite army from Beeroth; south and west was the army of Joshua. Out-manœuvred and out-generalled, they were in the most imminent danger of being caught between the two Israelite armies, and of being ground, like wheat, between the upper and nether millstones. They had no heart for further fight; the promise made to Joshua,—" there shall not a man of them stand before thee,"—was fulfilled; they broke and fled by the one way open to them, the way of the two Beth-horons.

372   THE ASTRONOMY OF THE BIBLE

Whilst this conjectural strategy attributes to Joshua a ready grasp of the essential features of the military position and skill in dealing with them, it certainly does not attribute to him any greater skill than it is reasonable to suppose he possessed. The Hebrews have repeatedly proved, not merely their valour in battle, but their mastery of the art of war, and, as Marcel Dieulafoy has recently shown,[1] the earliest general of whom we have record as introducing turning tactics in the field, is David in the battle of the valley of Rephaim, recorded in 2 Sam. v. 22–25 and 1 Chron. xiv. 13–17.

"The several evolutions of a complicated and hazardous nature which decided the fate of the battle would betoken, even at the present day, when successfully conducted, a consummate general, experienced lieutenants, troops well accustomed to manœuvres, mobile, and, above all, disciplined almost into unconsciousness, so contrary is it to our instincts not to meet peril face to face. . . . In point of fact, the Israelites had just effected in the face of the Philistines a turning and enveloping movement—that is to say, an operation of war considered to be one of the boldest, most skilful, and difficult attempted by forces similar in number to those of the Hebrews, but, at the same time, very efficacious and brilliant when successful. It was the favourite manœuvre of Frederick II, and the one on which his military reputation rests."

But though the Amorites had been discomfited by Joshua, they had not been completely surrounded; one way of escape was left open. More than this, it appears that they obtained a very ample start in the race along the north-western road. We infer this from the incident

[1] Marcel Dieulafoy, *David the King: an Historical Enquiry*, pp. 155–175.

of the hailstorm which fell upon them whilst rushing down the precipitous road between the Beth-horons; a storm so sudden and so violent that more of the Amorites died by the hailstones than had fallen in the contest at Gibeon. It does not appear that the Israelites suffered from the storm; they must consequently have, at the time, been much in the rear of their foes. Probably they were still "in the way that goeth up to Beth-horon"; that is to say, in the ascent some two miles long from Gibeon till the summit of the road is reached. There would be a special appropriateness in this case in the phrasing of the record that "the Lord discomfited the Amorites before Israel, and slew them with a great slaughter at Gibeon, and *chased* them along the way that goeth up to Beth-horon, and smote them to Azekah and unto Makkedah." There was no slaughter on the road between Gibeon and Beth-horon. It was a simple *chase;* a pursuit with the enemy far in advance.

The Israelites, general and soldiers alike, had done their best. The forced march all night up the steep ravines, the plan of the battle, and the way in which it had been carried out were alike admirable. Yet when the Israelites had done their best, and the heat and their long exertions had nearly overpowered them, Joshua was compelled to recognize that he had been but partly successful. He had relieved Gibeon; the Amorites were in headlong flight; he had cut them off from the direct road to safety, but he had failed in one most important point. He had not succeeded in surrounding them, and the greater portion of their force was escaping.

5.—The Miracle.

It was at this moment, when his scouts announced to him the frustration of his hopes, that Joshua in the anxiety lest the full fruits of his victory should be denied him, and in the supremest faith that the Lord God, in Whose hand are all the powers of the universe, was with him, exclaimed:

> "Sun, stand thou still upon Gibeon,
> And thou, Moon, in the valley of Ajalon!"

So his exclamation stands in our Authorized Version, but, as the marginal reading shows, the word translated "stand still" is more literally "be silent." There can be no doubt that this expression, so unusual in this connection, must have been employed with intention. What was it that Joshua is likely to have had in his mind when he thus spoke?

The common idea is that he simply wished for more time; for the day to be prolonged. But as we have seen, it was midday when he spoke, and he had full seven hours of daylight before him. There was a need which he must have felt more pressing. His men had now been seventeen hours on the march, for they had started at sunset— 7 p.m.—on the previous evening, and it was now noon the noon of a sub-tropical midsummer. They had marched at least twenty miles in the time, possibly considerably more according to the route which they had followed, and the march had been along the roughest of roads, and had included an ascent of 3400 feet—about the height of the summit of Snowdon above the sea-level. They must have

been weary, and have felt sorely the heat of the sun, now
blazing right overhead. Surely it requires no words to
labour this point. Joshua's one pressing need at that
moment was something to temper the fierce oppression of
the sun, and to refresh his men. This was what he prayed
for; this was what was granted him. For the moment
the sun seemed fighting on the side of his enemies, and
he bade it "Be silent." Instantly, in answer to his
command, a mighty rush of dark storm-clouds came
sweeping up from the sea.

Refreshed by the sudden coolness, the Israelites set out
at once in the pursuit of their enemies. It is probable
that for the first six miles they saw no trace of them, but
when they reached Beth-horon the Upper, and stood at
the top of its steep descent, they saw the Amorites again.
As it had been with their fathers at the Red Sea, when
the pillar of cloud had been a defence to them but the
means of discomfiture to the Egyptians, so now the storm-
clouds which had so revived them and restored their
their strength, had brought death and destruction to their
enemies. All down the rocky descent lay the wounded,
the dying, the dead. For "the Lord cast down great stones
from heaven upon them, unto Azekah, and they died: they
were more which died with hailstones than they whom the
children of Israel slew with the sword."

"The might of the Gentile, unsmote by the sword,
Had melted like snow in the glance of the Lord."

Far below them the panic-stricken remnants of the
Amorite host were fleeing for safety to the cities of

376   THE ASTRONOMY OF THE BIBLE

the Maritime Plain. The battle proper was over; the one duty left to the army of Israel was to overtake and destroy those remnants before they could gain shelter.

But the narrative continues. "The sun stayed in the midst of heaven, and hasted not to go down about a whole day." This statement evidently implies much more than the mere darkening of the sun by storm-clouds. For its interpretation we must return to the remaining incidents of the day.

These are soon told. Joshua pursued the Amorites to Makkedah, twenty-seven miles from Gibeon by the route taken. There the five kings had hidden themselves in a cave. A guard was placed to watch the cave; the Israelites continued the pursuit for an undefined distance farther; returned to Makkedah and took it by assault; brought the kings out of their cave, and hanged them.

"And it came to pass at the time of the going down of the sun, that Joshua commanded, and they took them down off the trees, and cast them into the cave wherein they had hidden themselves, and laid great stones on the mouth of the cave, unto this very day."

All these events—the pursuit for twenty-seven miles and more, the taking of Makkedah and the hanging of the kings—took place between noon and the going down of the sun, an interval whose normal length, for that latitude and at that time of the year, was about seven hours.

This is an abnormal feat. It is true that a single trained pedestrian might traverse the twenty-seven and odd miles, and still have time to take part in an assault on a town and to watch an execution. But it is an altogether different thing when we come to a large army. It is well known that the speed with which a body of men can move diminishes with the number. A company can march faster than a regiment; a regiment than a brigade; a brigade than an army corps. But for a large force *thirty miles in the entire day is heavy work.* " Thus Sir Archibald Hunter's division, in its march through Bechuanaland to the relief of Mafeking, starting at four in the morning, went on till seven or eight at night, covering as many as thirty miles a day at times." Joshua's achievement was a march fully as long as any of General Hunter's, but it was accomplished in less than seven hours instead of from fifteen to sixteen, and it followed straight on from a march seventeen hours in length which had ended in a battle. In all, between one sunset and the next he had marched between fifty and sixty miles besides fighting a battle and taking a town.

If we turn to the records of other battles fought in this neighbourhood, we find that they agree as closely as we could expect, not with Joshua's achievement, but with General Hunter's. In the case of the great victory secured by Jonathan, the gallant son of Saul, the Israelites smote the Philistines from Michmash to Ajalon;—not quite twenty miles. In the defeat of Cestius Gallus, the Jews followed him from Beth-horon to Antipatris, a little over twenty miles, the pursuit beginning at daybreak, and being evidently continued nearly till sundown. The pursuit of the Syrians under Nicanor by Judas Maccabæus seems also to have covered about the same distance, for

Nicanor was killed at the first onslaught and his troops took to flight.

It is not at all unusual to read in comments on the Book of Joshua that the "miracle" is simply the result of the dulness of the prose chronicler in accepting as literal fact an expression that originated in the poetic exuberance of an old bard. The latter, so it is urged, simply meaning to add a figure of dignity and importance to his song commemorating a great national victory, had written:—

"And the sun stood still, and the moon stayed,
  Until the nation had avenged themselves of their enemies,"

but with no more expectation that the stay of the moon would be accepted literally, than the singers, who welcomed David after the slaying of Goliath, imagined that any one would seriously suppose that Saul had actually with his own hand killed two thousand Philistines, and David twenty thousand. But, say they, the later prose chronicler, quoting from the ballad, and accepting a piece of poetic hyperbole as actual fact, reproduced the statement in his own words, and added, "the sun stayed in the midst of heaven, and hasted not to go down about a whole day."

Not so. The poem and the prose chronicle make one coherent whole. Working from the poem alone, treating the expressions in the first two lines merely as astronomical indications of time and place, and without the slightest reference to any miraculous interpretation, they lead to the inevitable conclusion that the time was noonday. This result certainly does not lie on the surface of the poem, and it was wholly beyond the power of the prose chronicler to have computed it, yet it is just in the supposed stupid gloss of the prose chronicler, and nowhere else, that we find this fact definitely stated: whilst the "miracle" recorded both by poem and prose narrative completely accords with the extraordinary distance traversed between noon and sunset.

Any man, however ignorant of science, if he be but careful and conscientious, can truthfully record an observation without any difficulty. But to successfully invent even the simplest astronomical observation requires very full knowledge, and is difficult even then. Every astronomer knows that there is hardly a single novelist, no matter how learned or painstaking, who can at this present day introduce a simple astronomical relation into his story, without falling into egregious error.

We are therefore quite sure that Joshua did use the words attributed to him; that the "moon" and "the valley of Ajalon" were not merely inserted in order to complete the parallelism by a bard putting a legend into poetic form. Nor was the prose narrative the result of an editor combining two or three narratives all written much after the date. The original records must have been made at the time.

All astronomers know well how absolutely essential it is to commit an observation to writing on the spot. Illustrations of this necessity could be made to any extent. One may suffice. In vol. ii. of the *Life of Sir Richard Burton*, by his wife, p. 244, Lady Burton says:—

"On the 6th December, 1882 . . . we were walking on the Karso (Opçona) alone; the sky was clear, and all of a sudden my niece said to me, 'Oh, look up, there is a star walking into the moon!' 'Glorious!' I answered. 'We are looking at the Transit of Venus, which crowds of scientists have gone to the end of the world to see.'"

The Transit of Venus did take place on December 6, 1882; and though Venus could have been seen without telescopic aid as a black spot on the sun's disc, nothing can be more unlike Venus in transit than "a star walking into the moon." The moon was not visible on that evening, and Venus was only visible when on the sun's disc, and appeared then, not as a star, but as a black dot.

No doubt Lady Burton's niece did make the exclamation attributed to her, but it must have been, not on December 6, 1882, but on some other occasion. Lady Burton may indeed have told her niece that this was the Transit of Venus, but that was simply because she did not know what a transit was, nor that it occurred in the daytime, not at night. Lady Burton's narrative was therefore not written at the time. So if the facts of the tenth chapter of Joshua, as we have it, had not been written at the time of the battle, some gross astronomical discordance would inevitably have crept in.

Let us suppose that the sun and moon did actually stand still in the sky for so long a time that between noon and sunset was equal to the full length of an ordinary day. What effect would have resulted that the Israelites could have perceived? This, and this only, that they would have marched twice as far between noon and sunset as they could have done in any ordinary afternoon. And this as we have seen, is exactly what they are recorded to have done.

The only measure of time, available to the Israelites, independent of the apparent motion of the sun, was the number of miles marched. Indeed, with the Babylonians, the same word (*kasbu*) was used to indicate three distinct, but related measures. It was a measure of time—the double hour; of celestial arc—the twelfth part of a great circle, thirty degrees, that is to say the space traversed by the sun in two hours; and it was a measure of distance on the surface of the earth—six or seven miles, or a two hours' march.

If, for the sake of illustration, we may suppose that the sun were to stand still for us, we should recognize it neither by sundial nor by shadow, but we should see that whereas our clocks had indicated that the sun had risen (we will say) at six in the morning, and had southed at twelve of noon; *it had not set until twelve of the night.* The register of work done, shown by all our clocks and watches, would be double for the afternoon what it had been for the morning. And if all our clocks and watches did thus register upon some occasion twice the interval between noon and sunset that they had registered between sunrise and noon, we should be justified in recording, as the writer of the book of Joshua has recorded, "The sun stood still in the midst of heaven, and hasted not to go down about a whole day."

The real difficulty to the understanding of this narrative has lain in the failure of commentators to put themselves back into the conditions of the Israelites. The Israelites had no time-measurers, could have had no time-measurers. A sundial, if any such were in existence, would only indicate the position of the sun, and therefore could give no evidence in the matter. Beside, a sundial is not a portable instrument, and Joshua and his men had something more pressing to do than to loiter round it. Clepsydræ or clocks are of later date, and no more than a sundial are they portable. Many comments, one might almost say most comments on the narrative, read as if the writers supposed that Joshua and his men carried stop-watches, and that their chief interest in the whole campaign was to see how fast the sun was moving. Since they had no such methods of measuring time, since it is not possible to suppose that over and above any material miracle that was wrought, the mental miracle was added of acquainting the Israelites for this occasion only with the Copernican system of astronomy, all that the words of the narrative can possibly mean is, that—

"the sun stood still in the midst of heaven, and hasted not to go down about a whole day,"

according to the only means which the Israelites had for testing the matter. In short, it simply states in other words, what, it is clear from other parts of the narrative, was actually the case, that the length of the march made between noon and sunset was equal to an ordinary march taking the whole of a day.

If we suppose—as has been generally done, and as it is quite legitimate to do, for all things are possible to God —that the miracle consisted in the slackening of the rotation of the earth, what effect would have been perceived by the Hebrews? This, and only this, that they would have accomplished a full day's march in the course of the afternoon. And what would have been the effects produced on all the neighbouring nations? Simply that they had managed to do more work than usual in the course of that afternoon, and that they felt more than usually tired and hungry in the evening.

But would it have helped the Israelites for the day to have been thus actually lengthened? Scarcely so, unless they had been, at the same time, endowed with supernatural, or at all events, with unusual strength. The Israelites had already been 31 hours without sleep or rest, they had made a remarkable march, their enemies had several miles start of them; would not a longer day have simply given the latter a better chance to make good their flight, unless the Israelites were enabled to pursue them with unusual speed? And if the Israelites were so enabled, then no further miracle is required; for them the sun would have "hasted not to go down about a whole day."

Leaving the question as to whether the sun appeared to stand still through the temporary arrest of the earth's rotation, or through some exaltation of the physical powers of the Israelites, it seems clear, from the foregoing analysis of the narrative, that both the prose account and the poem were written by eye-witnesses, who recorded what they had themselves seen and heard whilst every detail was fresh in their memory. Simple as the astronomical references are, they are very stringent, and can only have been supplied by those who were actually present.

Nothing can be more unlike poetic hyperbole than the sum of actual miles marched to the men who trod them; and these very concrete miles were the gauge of the lapse of time. For just as "nail," and "span," and "foot," and "cubit," and "pace" were the early measures of small distance, so the average day's march was the early measure of long distance. The human frame, in its proportions and in its abilities, is sufficiently uniform to have furnished the primitive standards of length. But the relation established between time and distance as in the case of a day's march, works either way, and is employed in either direction, even at the present day. When the Israelites at the end of their campaign returned from Makkedah to Gibeon, and found the march, though wholly unobstructed, was still a heavy performance for the whole of a long day, what could they think, how could they express themselves, concerning that same march made between noon and sundown? Whatever construction we put upon the incident, whatever explanation we may offer for it, to all the men of Israel, judging the events of the afternoon by the only standard within their reach, the eminently practical standard of the miles they had marched, the only conclusion at which they could arrive was the one they so justly drew—

"The sun stayed in the midst of heaven and hasted not to go down about a whole day. And there was no day like that before it or after it, that the Lord hearkened unto the voice of a man: for the Lord fought for Israel."

CHAPTER II

THE DIAL OF AHAZ

THE second astronomical marvel recorded in the Scripture narrative is the going back of the shadow on the dial of Ahaz, at the time of Hezekiah's recovery from his dangerous illness.

It was shortly after the deliverance of the kingdom of Judah from the danger threatened it by Sennacherib king of Assyria, that Hezekiah fell "sick unto death." But in answer to his prayer, Isaiah was sent to tell him—

"Thus saith the Lord, the God of David thy father, I have heard thy prayer, I have seen thy tears: behold, I will heal thee: on the third day thou shalt go up unto the house of the Lord. And I will add unto thy days fifteen years; and I will deliver thee and this city out of the hand of the king of Assyria; and I will defend this city for Mine own sake, and for My servant David's sake. And Isaiah said, Take a lump of figs. And they took and laid it on the boil, and he recovered. And Hezekiah said unto Isaiah, What shall be the sign that the Lord will heal me, and that I shall go up into the house of the Lord the third day? And Isaiah said, This sign shalt thou have of the Lord, that the Lord will do the thing that He hath spoken: shall the shadow go forward ten degrees, or go back ten degrees? And Hezekiah answered, It is a light thing for the shadow to go down ten degrees:

nay, but let the shadow return backward ten degrees. And Isaiah the prophet cried unto the Lord: and He brought the shadow ten degrees backward, by which it had gone down in the dial of Ahaz."

The narrative in the Book of Isaiah gives the concluding words in the form—

"So the sun returned ten degrees, by which degrees it was gone down."

The narrative is complete as a record of the healing of king Hezekiah and of the sign given to him to assure him that he should recover; complete for all the ordinary purposes of a narrative, and for readers in general. But for any purpose of astronomical analysis the narrative is deficient, and it must be frankly confessed that it does not lie within the power of astronomy to make any use of it.

It has been generally assumed that it was an actual sundial upon which this sign was seen. We do not know how far back the art of dialling goes. The simplest form of dial is an obelisk on a flat pavement, but it has the very important drawback that the graduation is different for different times of the year. In a properly constructed dial the edge of the style casting the shadow should be made parallel to the axis of the earth. Consequently a dial for one latitude is not available without alteration when transferred to another latitude. Some fine types of dials on a large scale exist in the observatories built by Jai Singh. The first of these—that at Delhi—was probably completed about 1710 A.D. They are, therefore, quite modern, but afford good illustrations of the type of structure which we can readily conceive of as having been built in what has been termed the Stone Age of astronomy. The principal of these buildings, the Samrat Yantra, is a long staircase in the meridian leading up to nothing, the shadow falling on to a great semicircular arc which it crosses. The slope of the staircase is, of course, parallel to the earth's axis.

It has been suggested that if such a dial were erected at Jerusalem, and the style were that for a tropical latitude, at certain times of the year the shadow would appear to go backward for a short time. Others, again, have suggested that if a small portable dial were tilted the same phenomenon would show itself. It is, of course, evident that no such suggestion at all accords with the narrative. Hezekiah was now in the fourteenth year of his reign, the dial—if dial it was—was made by his father, and the "miracle" would have been reproduced day by day for a considerable part of each year, and after the event it would have been apparent to every one that the "miracle" continued to be reproduced. If this had been the case, it would say very little for the astronomical science of the wise men of Merodach-Baladan that he should have sent all the way from Babylon to Jerusalem "to inquire of the wonder that was done in the land" if the wonder was nothing more than a wrongly mounted dial.

Others have hazarded the extreme hypothesis, that there might have been an earthquake at the time which dipped the dial in the proper direction, and then restored it to its proper place; presumably, of course, without doing harm to Jerusalem, or any of its buildings, and passing unnoticed by both king and people.

A much more ingenious theory than any of these was communicated by the late J. W. Bosanquet to the Royal Asiatic Society in 1854. An eclipse of the sun took place on January 11, 689 B.C. It was an annular eclipse in Asia Minor, and a very large partial eclipse at Jerusalem, the greatest phase taking place nearly at local noon. Mr. Bosanquet considers that the effect of the partial eclipse would be to practically shift the centre of the bright body casting the shadow. At the beginning of the annular phase, the part of the sun uncovered would be a crescent in a nearly vertical position; at mid eclipse the crescent would be in a horizontal position; at the end of the annular phase the crescent would again be in a vertical position; so that the exposed part of the sun would appear to move down and up in the sky over a very small distance. It is extremely doubtful whether any perceptible effect could be so produced on the shadow, and one wholly fails to understand why the eclipse itself should not have been given as the sign, and why neither the king nor the people seem to have noticed that it was in progress. It is, however, sufficient to say that modern chronology shows that Hezekiah died ten years before the eclipse in question, so that it fell a quarter of a century too late for the purpose, and no other eclipse is available to take its place during the lifetime of Hezekiah.

But there is no reason to think that the word rendered in our Authorized Version as "dial" was a sundial at all,

THE DIAL OF AHAZ 389

The word translated "dial" is the same which is also rendered "degrees" in the A.V. and "steps" in the R.V., as is shown in the margin of the latter. It occurs in the prophecy of Amos, where it is rendered "stories" or "ascensions." It means an "ascent," a "going up," a "step." Thus king Solomon's throne had six *steps*, and there are fifteen Psalms (cxx.-cxxxiv). —that are called "songs of degrees," that is "songs of steps."

We do not know how the staircase of Ahaz faced, but we can form some rough idea from the known positions of the Temple and of the city of David, and one or two little hints given us in the narrative itself. It will be noted that Hezekiah uses the movement of the shadow downward, as equivalent to its going forward. The going forward of course meant its ordinary direction of motion at that time of day; so the return of the shadow backward meant that the shadow went up ten steps, for in the Book of Isaiah it speaks of the sun returning "ten degrees by which degrees it was gone down." It was therefore in the afternoon, and the sun was declining, when the sign took place. It is clear, therefore, that the staircase was so placed that the shadow went down the stairs as the sun declined in the sky. The staircase, therefore, probably faced east or north-east, as it would naturally do if it led from the palace towards the Temple. No doubt there was a causeway at the foot of this staircase, and a corresponding ascent up the Temple hill on the opposite side of the valley.

We can now conjecturally reproduce the circumstances.

It was afternoon, and the palace had already cast the upper steps of the staircase into shadow. The sick king, looking longingly towards the Temple, could see the lower steps still gleaming in the bright Judean sunshine. It was natural therefore for him to say, when the prophet Isaiah offered him his choice of a sign, "Shall the shadow go forward ten steps, or back ten steps?" that it was "a light thing for the shadow to go down ten steps: nay, but let the shadow return backward ten steps." It would be quite obvious to him that a small cloud, suitably placed, might throw ten additional steps into shadow.

It will be seen that we are left with several details undetermined. For the staircase, wherever constructed, was probably not meant to act as a sundial, and was only so used because it chanced to have some rough suitability for the purpose. In this case the shadow will probably have been thrown, not by a properly constructed gnomon, but by some building in the neighbourhood. And as we have no record of the direction of the staircase, its angle of inclination, its height, and the position of the buildings which might have cast a shadow upon it, we are without any indication to guide us.

When the queen of Sheba came to visit king Solomon, and saw all his magnificence, one of the things which specially impressed her was "his ascent by which he went up unto the house of the Lord." This was "the causeway of the going up," as it is called in the First Book of Chronicles. We are told of a number of alterations, made in the Temple furniture and buildings by king Ahaz, and it is said that "the covered way for the sabbath that they had built in the house, and the king's entry without, turned he unto (*margin*, round) the house of the Lord, because of the king of Assyria." That is to say, Ahaz considered that Solomon's staircase was too much exposed in the case of a siege, being without the Temple enclosure. This probably necessitated the construction of a new staircase, which would naturally be called the staircase of Ahaz. That there was, in later times, such a staircase at about this place we know from the route taken by the triumphal procession at the time of the dedication of the wall of Jerusalem under Nehemiah :—

"At the fountain gate, which was over against them, they went up by the stairs of the City of David, at the going up of the wall, above the house of David, even unto the water gate eastward."

In this case there would be a special appropriateness in the sign that was offered to Hezekiah. The sign that he would be so restored, as once again to go up to the house of the Lord, was to be given him on the very staircase by which he would go. He was now thirty-eight years old, and had doubtless watched the shadow of the palace descend the staircase in the afternoon, hundreds of times; quite possibly he had actually seen a cloud make the shadow race forward. But the reverse he had never seen. Once a step had passed into the shadow of the palace, it did not again emerge until the next morning dawned.

The sign then was this: It was afternoon, probably approaching the time of the evening prayer, and the court officials and palace attendants were moving down the staircase in the shadow, when, as the sick king watched them from above, the shadow of the palace was rolled back up the staircase, and a flood of light poured down on ten of the broad steps upon which the sun had already set. How this lighting of the ten steps was brought about we are not told, nor is any clue given us on which we can base a conjecture. But this return of light was a figure of what was actually happening in the life of the king himself. He had already, as it were, passed into the shadow that only deepens into night. As he sang himself after his recovery—

"I said, In the noontide of my days I shall go into the gates
    of the grave :
I am deprived of the residue of my years.'
I said, I shall not see the Lord, even the Lord in the land
    of the living :
I shall behold man no more with the inhabitants of the world."

But now the light had been brought back to him, and he could say—

"The living, the living, he shall praise Thee, as I do this day:
The father to the children shall make known Thy truth.
The Lord is ready to save me :
Therefore we will sing my songs to the stringed instruments
All the days of our life in the house of the Lord."

CHAPTER III

THE STAR OF BETHLEHEM

No narrative of Holy Scripture is more familiar to us than that of the visit of the wise men from the East to see Him that was born King of the Jews. It was towards the end of the reign of Herod the Great that they arrived at Jerusalem, and threw Herod the king and all the city into great excitement by their question—

"Where is He that is born King of the Jews? For we have seen His star in the east, and are come to worship Him."

Herod at once gathered all the chief priests and scribes of the people together, and demanded of them where the Messiah should be born. Their reply was distinct and unhesitating—

"In Bethlehem of Judæa: for thus it is written by the Prophet, And thou Bethlehem, in the land of Juda, art not the least among the princes of Juda: for out of thee shall come a Governor, that shall rule My people Israel. Then Herod, when he had privily called the wise men, inquired of them diligently what time the star appeared. And he sent them to Bethlehem, and said, Go and search diligently for the young Child; and when ye have found Him, bring me word again, that I may come and worship Him also. When they had heard the king, they departed; and, lo, the star, which they saw in the east, went before them, till it came and stood over where the young Child was. When they saw the star, they rejoiced with exceeding great joy."

So much, and no more are we told of the star of Bethlehem, and the story is as significant in its omissions as in that which it tells us.

What sort of a star it was that led the wise men; how they learnt from it that the King of the Jews was born; how it went before them; how it stood over where the young Child was, we do not know. Nor is it of the least importance that we should know. One verse more, and that a short one, would have answered these inquiries; it would have told us whether it was some conjunction of the planets; whether perchance it was a comet, or a "new" or "temporary" star; or whether it was a supernatural light, like the pillar of fire that guided the children of Israel in the wilderness. But that verse has not been given. The twelve or twenty additional words, which could have cleared up the matter, have been withheld, and there can be no doubt as to the reason. The "star," whatever its physical nature, was of no importance, except as a guide to the birthplace of the infant Jesus. Information about it would have drawn attention from the object of the narrative; it would have given to a mere sign-post the importance which belonged only to "the Word made flesh."

We are often told that the Bible should be studied precisely as any other book is studied. Yet before we can criticize any book, we must first ascertain what was the purpose that the author had in writing it. The history of England, for instance, has been written by many persons and from many points of view. One man has traced the succession of the dynasties, the relationships of the successive royal families, and the effect of the administrations of the various kings. Another has chiefly considered the development of representative government and of parliamentary institutions. A third has concerned himself more with the different races that, by their fusion, have formed the nation as it is to-day. A fourth has dealt with the social condition of the people, the increase of comfort and luxury. To a fifth the true history of England is the story of its expansion, the foundation and growth of its colonial empire. While to a sixth, its religious history is the one that claims most attention, and the struggles with Rome, the rise and decay of Puritanism, and the development of modern thought will fill his pages. Each of these six will select just those facts, and those facts only, that are relevant to his subject. The introduction of irrelevant facts would be felt to mark the ignorant or unskilful workman. The master of his craft will keep in the background the details that have no bearing on his main purpose, and to those which have but a slight bearing he will give only such notice as their importance in this connection warrants.

The purpose of the Bible is to reveal God to us, and to teach us of our relationship to Him. It was not intended to gratify that natural and laudable curiosity which has been the foundation of the physical sciences. Our own efforts, our own intelligence can help us here, and the Scriptures have not been given us in order to save us the trouble of exerting them.

There is no reason for surprise, then, that the information given us concerning the star is, astronomically, so imperfect. We are, indeed, told but two facts concerning it. First that its appearance, in some way or other, informed the wise men, not of the birth of *a* king of the Jews, but of *the* King of the Jews, for Whose coming not Israel only, but more or less consciously the whole civilized world, was waiting. Next, having come to Judæa in consequence of this information, the "star" pointed out to them the actual spot where the new-born King was to be found. "It went before them till it came and stood over where the young Child was." It may also be inferred from Matt. ii. 10 that in some way or other the wise men had for a time lost sight of the star, so that the two facts mentioned of it relate to two separate appearances. The first appearance induced them to leave the East, and set out for Judæa; the second pointed out to them the place at Bethlehem where the object of their search was to be found. Nothing is told us respecting the star except its work as a guide.

Some three centuries ago the ingenious and devout Kepler supposed that he could identify the Star with a conjunction of the planets Jupiter and Saturn, in the constellation Pisces. This conjunction took place in the month of May, B.C. 7, not very long before the birth of our Lord is supposed to have taken place.

But the late Prof. C. Pritchard has shown, first, that a similar and closer conjunction occurred 59 years earlier, and should therefore have brought a Magian deputation to Judæa then. Next, that the two planets never approached each other nearer than twice the apparent diameter of the moon, so that they would have appeared, not as one star, but as two. And thirdly, if the planets had seemed to stand over Bethlehem as the wise men left Jerusalem, they most assuredly would not have appeared to do so when they arrived at the little city. Ingenious as the suggestion was, it may be dismissed as unworthy of serious consideration.

Another suggestion shows upon what slight foundations a well-rounded legend may be built. In the year 1572 a wonderful "new star" appeared in the constellation Cassiopeia. At its brightest it outshone Venus, and, though it gradually declined in splendour, it remained visible for some sixteen months. There have been other instances of outbursts of bright short-lived stars; and brief notices, in the annals of the years 1265 and 952 may have referred to such objects, but more probably these were comets. The guess was hazarded that these objects might be one and the same; that the star in Cassiopeia might be a "variable" star, bursting into brilliancy about every 315 or 316 years; that it was the star that announced the birth of our Lord, and that it would reappear towards the end of the nineteenth century to announce His second coming.

One thing more was lacking to make the legend complete, and this was supplied by the planet Venus, which shines with extraordinary brilliance when in particular parts of her orbit. On one of these occasions, when she was seen as a morning star in the east, some hazy recollection of the legend just noticed caused a number of people to hail her as none other than the star of Bethlehem at its predicted return.

There is no reason to suppose that the star of 1572 had ever appeared before that date, or will ever appear again. But in any case we are perfectly sure that it could not have been the star of Bethlehem. For Cassiopeia is a northern constellation, and the wise men, when they set out from Jerusalem to Bethlehem must have had Cassiopeia and all her stars behind them.

The fact that the "star" went before them and stood over where the young Child lay, gives the impression that it was some light, like the Shekinah glory resting on the Ark in the tabernacle, or the pillar of fire which led the children of Israel through the wilderness. But this view raises the questions as to the form in which it first appeared to the wise men when they were still in the East, and how they came to call it a star, when they must have recognized how very unstarlike it was. Whilst, if what they saw when in the East was really a star, it seems most difficult to understand how it can have appeared to go before them and to stand over the place where the young Child lay.

I have somewhere come across a legend which may possibly afford the clue, but I have not been able to find that the legend rests upon any authority. It is that the star had been lost in the daylight by the time that the

THE STAR OF BETHLEHEM 399

wise men reached Jerusalem. It was therefore an evening star during their journey thither. But it is said that when they reached Bethlehem, apparently nearly at midday, one of them went to the well of the inn, in order to draw water. Looking down into the well, he saw the star, reflected from the surface of the water. This would of course be an intimation to them that the star was directly overhead, and its re-observation, under such unusual circumstances, would be a sufficient assurance that they had reached the right spot. Inquiry in the inn would lead to a knowledge of the visit of the shepherds, and of the angelic message which had told them where to find the Babe born in the city of David, " a Saviour, Which is Christ the Lord."

If this story be true, the "Star of Bethlehem" was probably a "new star," like that of 1572. Its first appearance would then have caused the Magi to set out on their journey, though it does not appear how they knew what it signified, unless we suppose that they were informed of it in a dream, just as they were afterwards warned of God not to return to Herod. Whilst they were travelling the course of the year would bring the star, which shone straight before them in the west after sunset every evening, nearer and nearer to the sun. We may suppose that, like other new stars, it gradually faded, so that by the time the wise men had reached Jerusalem they had lost sight of it altogether. Having thus lost it, they would probably not think of looking for it by daylight, for it is no easy thing to detect by daylight even Venus at her greatest brilliancy, unless one knows exactly where to look.

The difficulty does not lie in any want of brightness, but in picking up and holding steadily so minute a point of light in the broad expanse of the gleaming sky. This difficulty would be overcome for them, according to this story, by the well, which acted like a tube to direct them exactly to the star, and like a telescope, to lessen the sky glare. It would be also necessary to suppose that the star was flashing out again with renewed brilliancy. Such a brief recovery of light has not been unknown in the case of some of our "new" or "temporary" stars.

I give the above story for what it is worth, but I attach no importance to it myself. Some, however, may feel that it removes what they had felt as a difficulty in the narrative,—namely, to understand how the star could "stand over where the young Child lay." It would also explain, what seems to be implied in the narrative, how it happened that the Magi alone, and not the Jews in general, perceived the star at its second appearance.

For myself, the narrative appears to me astronomically too incomplete for any astronomical conclusions to be drawn from it. The reticence of the narrative on all points, except those directly relating to our Lord Himself, is an illustration of the truth that the Scriptures were not written to instruct us in astronomy, or in any of the physical sciences, but that we might have eternal life.

"And this is life eternal, that they might know Thee the Only true God, and Jesus Christ, whom Thou hast sent."

A TABLE OF SCRIPTURAL REFERENCES

Page.	Book.	Chap. and Verse.	Page.	Book.	Chap. and Verse.
9	I. Kings	iv. 29-34	45	Jer.	x. 18 (R.V.)
10	Wisdom	vii. 17-22 (R.V.)	,,	Psalm	cxxxv. 7
11	Psalm	viii. 3, 4	46	Job	xxxvii. 16
15	Eccl.	i. 9	49	,,	xxxvi. 29
17	Gen.	i. 1	,,	Gen.	vii. 11
,,	I. Chron.	xvi. 26	50	II. Kings	vii. 1, 2
,,	Deut.	vi. 4	,,	Mal.	iii. 10
,,	Mark	xii. 29	,,	Hos.	vi. 4
,,	Neh.	ix. 6	,,	Dan.	viii. 8
19	Heb.	xi. 3 5	,,	Ezek.	xxxvii. 9
20	II. Pet.	iii. 8	51	Jer.	xlix. 36
: 2	Psalm	cxi. 2-4 (R.V.)	,,	Eccl.	i. 6
,,	Gen.	ii. 3	,,	Isaiah	xi. 12
23	Exod.	xx. 10, 11	,,	,,	xl. 22
,,	,,	xxxi. 16, 17	,,	Prov.	viii. 27
,,	Gen.	i. 14	52	Job	xxii. 14 (R.V. margin)
25	,,	i. 1			
32	Exod.	xv. 4, 5	,,	,,	xxxvi. 10 (R.V.)
35	Gen.	i. 6-8	,,	Gen.	i. 9
,,	,,	i. 14	,,	Psalm	xxiv. 2
36	,,	i. 20	,,	,,	cxxxvi. 6
,,	Job	xxxvii. 18 (R.V.)	53	Ezek.	xxxi. 4
,,	Num.	xvii. 39	,,	Gen.	vii. 11
,,	Isaiah	xl. 19	,,	,,	viii. 2
37	Jer.	x. 9	,,	Job	xxxviii. 16
,,	Psalm	cxxxvi. 6	,,	Prov.	iii. 20
38	Heb.	i. 3	,,	Jer.	v. 22
39	II. Sam.	xxii. 8	54	Job	xxxviii. 8
,,	Job	xxvi. 11	,,	Prov.	viii. 27, 29
,,	,,	xxvi. 7	55	Josh.	x. 13
,,	I. Sam.	ii. 8	,,	Psalm	xix. 1-6 (R.V.)
40	Psalm	lxxv. 3	56	I. Kings	xxii. 19
,,	,,	civ. 2	57	Jer.	xxxiii. 22
,,	Isaiah	xl. 22	,,	Deut.	iv. 15, 19
,,	Amos	ix. 6	58	Job	xxxviii. 7
,,	Num.	xxxiv. 4	,,	Judges	v. 20
,,	II. Sam.	xv. 30	,,	II. Kings	vi. 14-17
41	Psalm	cxlviii. 4	60	Job	xxxviii. 32 (R.V.)
,,	Song of Three Children	38	,,	Psalm	cxi. 2
,,	Amos	v. 8	,,	Rev.	ii. 26, 28
,,	,,	ix. 6	61	Isaiah	xiv. 12-14
,,	Eccl.	i. 7	,,	Rev.	xxii. 16
,,	Isaiah	lv. 10 (R.V.)	62	Jer.	xxxi. 36
42	,,	lv. 11	,,	Isaiah	xl. 26-31 (R.V.)
,,	Job	xxxvi. 26-28 (R.V.)	63	Gen.	i. 14-19
			64	Deut.	xxxiii. 14 (R.V.)
,,	Judges	v. 4	,,	I. John	i. 5
,,	Psalm	lxxvii. 17	,,	Psalm	xxvii. 1
,,	,,	cxlvii. 8	,,	Isaiah	lx. 19
,,	Prov.	xvi. 15	,,	John	i. 9
,,	Eccl.	xii. 2	,,	Psalm	lxxxiv. 11
,,	Isaiah	v. 6	,,	Mal.	iv. 2
,,	Jude	12	65	James	i. 17
,,	Nahum	i. 3	,,	Psalm	cxxxix. 12
,,	Isaiah	xviii. 4	,,	Deut.	iv. 19
43	Eccl.	xi. 3			xvii. 2, 3
,,	Job	xxvi. 8	66	II. Kings	xxiii. 11
,,	,,	xxxviii. 34-37	,,	Ezek.	viii. 16
44	,,	xxxviii. 19-29 (R.V.)	,,	,,	viii. 16
			,,	Job	xxxi. 26
,,	Psalm	xviii. 6-17 (R.V.)	,,	Cant.	vi. 10
			67	Judges	viii. 13

TABLE OF SCRIPTURAL REFERENCES

Page.	Book.	Chap. and Verse.	Page.	Book.	Chap. and Verse.
67	Judges	xiv. 18	89	Jer.	vii. 18
,,	Jer.	xliii. 18	90	,,	xliv. 17, 18
68	Isaiah	xix. 18	91	Isaiah	xxx. 26
,,	Cant.	vi. 10	,,	,,	lx. 20
69	Psalm	lxxii. 5	92	Psalm	cxxi. 6
,,	,,	lxxii. 17	,,	,,	civ. 19-24 (R.V.)
70	,,	lxxxix. 36	96	Gen.	xv. 5
,,	,,	l. 1	97	Psalm	cxlvii. 4
,,	,,	cxiii. 3	,,	Isaiah	xl. 22
,,	,,	xix. 6	98	I. Cor.	xv. 41
,,	Eccl.	i. 3	99	Prov.	xxv. 3
71	Job	xxxviii. 12-14	,,	Job	xi. 7, 8
,,	,,	xxxviii. 14 (R.V.)	,,	,,	xxii. 12
,,	Eccl.	i. 5	,,	Jer.	xxxi. 37
72	Job	xxvi. 7	100	Psalm	ciii. 11, 12
,,	Psalm	xix. 6	107	Joel	ii. 30
,,	II. Kings	iv. 19	,,	Gen.	iii. 24
,,	Psalm	cxxi. 6	,,	Heb.	i. 7
,,	Isaiah	xlix. 10	,,	I. Chron.	xxi. 16
,,	Rev.	vii. 16	108	Jude	13
,,	Deut.	xxxiii. 14	113	Acts	xix. 35 (R.V.)
73	James	i. 17	116	Rev.	vi. 13
,,	Job	xxxviii. 33	,,	Isaiah	xxxiv. 4
,,	Wisdom	vii. 18	,,	Rev.	viii. 10
78	Rom.	i. 20-23	,,	Jude	13
79	John	ix. 4	117	Job	iii. 9 (margin)
80	Psalm	lxxxi. 3	,,	,,	xli. 18
,,	Prov.	vii. 20	,,	,,	xxxvii. 22 (R.V.)
82	Isaiah	lx. 20	119	Jer.	x. 2
83	Num.	x. 10	122	Wisdom	vii. 18
,,	Psalm	lxxxi. 3	123	Amos	i. 1
,,	Isaiah	iii. 18	,,	Zech.	xiv. 5
84	Gen.	xxxvii. 9	,,	Gen.	i. 14
,,	Jer.	viii. 2	124	Joel	ii. 10
,,	Psalm	civ. 19	,,	,,	ii. 30, 31
,,	,,	lxxxix. 36, 37	,,	Acts	ii. 19, 20
,,	,,	cxxxvi. 9	,,	Rev.	vi. 12
,,	Jer.	xxxi. 35	,,	Amos	viii. 9
,,	Eccl.	xii. 2	125	Micah	iii. 6
,,	Isaiah	xiii. 10	,,	Isaiah	xiii. 10
,,	Ezek.	xxxii. 7	,,	Jer.	xv. 9
,,	Joel	ii. 10, 31	,,	Ezek.	xxxii. 7, 8
,,	,,	iii. 15	129	Mal.	iv. 2
,,	Hab.	iii. 11	,,	James	i. 17 (R.V.)
,,	Exod.	ii. 2	131	Gen.	xiv. 5
85	Deut.	xxxiii. 13, 14	132	Isaiah	xlvi. 1
,,	II. Kings	xv. 13	,,	,,	xiv. 12
,,	Dan.	iv. 29	,,	II. Peter	i. 19
,,	Ezra	vi. 15	,,	Isaiah	lxv. 11
,,	Neh.	i. 1	,,	Dan.	v. 26 (R.V.)
,,	I. Kings	vi. 1, 37, 38	133	Amos	v. 25, 26
,,	,,	viii. 2	,,	Acts	vii. 43
,,	Cant.	vi. 10	143	Isaiah	viii. 10
,,	Isaiah	lxxiv. 23	144	Ezek.	xxi. 21 (R.V.)
86	,,	xxx. 26	,,	Isaiah	xlvii. 12, 13
,,	Rev.	xix. 6-8	,,	Jer.	x. 2
,,	Gen.	xxxvii. 9	150	Acts	xvii. 24-28
87	,,	xxxvii. 10	163	Gen.	ix. 13
,,	Job	xxxi. 26-28 (R.V.)	164	,,	iii. 15
88	Deut.	iv. 12, 15, 16, 19	166	,,	iii. 24
,,	Judges	viii. 21	,,	Ezek.	i. 5
,,	Isaiah	iii. 18	,,	Rev.	iv. 7 (R.V.)
,,	II. Kings	xxiii. 13	,,	Ezek.	x. 20
89	Gen.	xiv. 5	,,	I. Kings	vi. 29, 32
,,	I. Sam.	xxxi. 10	167	Gen.	x. 9
,,	II. Kings	xxiii. 13	169	Psalm	lxxx. 1

TABLE OF SCRIPTURAL REFERENCES 403

Page.	Book.	Chap. and Verse.	Page.	Book.	Chap. and Verse.
173	Gen.	vi. 19	239	Amos	v. 8
,,	,,	vii. 2	241	Job	xxxviii. 26
184	Psalm	l. 5	242	,,	xxvi. 13
186	Gen.	xxxvii. 9	,,	Isaiah	xlv. 7
189	,,	xlix. 9	243	Job	xxxviii. 32
,,	Rev.	v. 5	251	,,	xxxviii. 32
190	Deut.	xxxiii. 17 (R.V.)	,,	II. Kings	xxiii. 5
,,	Gen.	xlix. 6 (R.V.)	,,	Deut.	iv. 19
,,	,,	xlix. 4, 17	,,	Job	ix. 9
191	Num.	xxiii. 7, 24 (R.V.)	,,	,,	xxxviii. 31, 32
,,	,,	xxiv. 9 (R.V.)	,,	,,	xxxvii. 9
,,	,,	xxiv. 8 (R.V.)	252	Exod.	xxxii.
,,	,,	xxiv. 7 (R.V.)	,,	I. Kings	xii.
193	Exod.	xxxii. 1	254	II. Kings	xxiii. 5
,,	Acts	vii. 41, 42	,,	Job	ix., xxxviii.
,,	Exod.	xx. 8	257	,,	xxxviii. 33
,,	,,	xx. 4, 5	,,	Luke	xi. 2
194	Deut.	iv. 15	258	Job	ix. 9
,,	Psalm	cvi. 20	,,	,,	xxxviii. 31-33
,,	Acts	vii. 42	259	,,	xxxvii. 9
,,	I. Kings	xii. 28	260	Isaiah	l. 9
195	Rev.	v. 5	262	Job	xxxvii. 9
203	Job	iii. 8, 9 (R.V.)	271	Gen.	i. 14
,,	,,	xli.	,,	Deut.	iv. 19
,,	Psalm	civ. 25	273	Exod.	xii. 18, 19
,,	Isaiah	xxvii. 1	,,	Lev.	xxiii. 32
204	Job	xxvi. 12, 13	275	Psalm	lv. 17
205	Isaiah	xxx. 7 (R.V.)	,,	Job	iii. 9 (margin)
,,	,,	li. 9, 10 (R.V.)	,,	Cant.	ii. 17
,,	Psalm	lxxxix. 9, 10	,,	Gen.	xxxii. 24, 26
206	Ezek.	xxxii. 2 (R.V.)	,,	Josh.	vi. 15
,,	Rev.	xx. 2	,,	Judges	xix. 25
,,	Ezek.	xxxii. 4 (R.V.)	,,	II. Sam.	ii. 32
,,	,,	xxix. 3, 5	276	Gen.	xxxii. 31
207	Rev.	xii. 6 (R.V.)	,,	Exod.	xvi. 21
208	,,	xii. 15, 16 (R.V.)	,,	I. Sam.	xi. 9
209	Job	iii. 9 (R.V.)	,,	II. Sam.	iv. 5
,,	,,	xli. 18 (R.V.)	,,	I. Kings	xviii. 26
210	Psalm	xix. 5	,,	Judges	xix. 8, 9
211	I. Kings	xviii. 27	,,	Job	vii. 2
,,	Isaiah	xxx. 31	,,	Jer.	vi. 4
212	Psalm	lxxiv. 12-17	,,	Prov.	vii. 9
215	Job	ix. 9	277	Exod.	xii. 6
,,	,,	xxxviii. 31	,,	,,	xvi. 12
,,	Amos	v. 8	,,	,,	xxx. 8
217	Isaiah	lxv. 11	,,	Levit.	xxiii. 5
218	II. Kings	xvii. 30	,,	Num.	ix. 3
,,	Gen.	xlix. 22	,,	,,	xxviii. 4
220	Rev.	i. 12, 13, 15	278	Deut.	xvi. 6
,,	,,	i. 20	279	Exod.	xxx. 8
223	I. Peter	iii. 20	280	I. Cor.	xv. 52
,,	Amos	v. 8	,,	Psalm	lxiii. 6
,,	Job	xxxviii. 31	,,	,,	cxix. 148
224	Cant.	ii. 11-13 (R.V.)	,,	Lam.	ii. 19
225	Job	xxxviii. 4	281	Judges	vii. 19
,,	,,	xxxviii. 31 (R.V.)	,,	Exod.	xiv. 24
231	,,	ix. 9	,,	I. Sam.	xi. 11
,,	,,	xxxviii. 31	,,	Matt.	xiv. 25
,,	Amos	v. 8	,,	Mark	vi. 48
,,	Isaiah	xiii. 10	,,	Dan.	iii. 6, 15
,,	Prov.	i. 22	,,	,,	iv. 19, 33
234	Gen.	x. 8	,,	,,	v. 5
235	,,	x. 10	,,	Job	xxxviii. 12
238	Isaiah	xiv. 13, 14	282	Acts	i. 12
239	,,	xiii. 9-11	,,	Matt.	xx.

404 TABLE OF SCRIPTURAL REFERENCES

Page.	Book.	Chap. and Verse.	Page.	Book.	Chap. and Verse.
282	John	xi. 9, 10	327	Lev.	xxvi. 2, 21
291	Exod.	xx. 11	,,	,,	xxvi. 38–35
,,	Psalm	cxviii. 24	,,	Deut.	xv. 1
293	II. Kings	iv. 23	328	,,	xxxi. 10, 11
,,	Isaiah	i. 13, 14	,,	Jer.	xxxiv.
294	Isaiah	lxvi. 23	,,	Lev.	xxvi. 32–35
,,	Amos	viii. 5	,,	II. Chron.	xxxvi. 21
,,	Col.	ii. 16	329	Neh.	x. 31
,,	Num.	xxviii.	,,	Lev.	xxv. 8–10
,,	I. Chron.	xxiii.	330	Num.	xxxvi. 4
,,	II. Chron.	ii.	,,	Isaiah	lxi. 2
,,	,,	xxix.	,,	Ezek.	xlvi. 17
,,	Ezek.	xlv.	332	Lev.	xxv. 8, 10
,,	Ezra	iii.	,,	,,	xxv. 11, 12
,,	Neh.	x.	333	,,	xxv. 22
295	Num.	xxix. 1	,,	,,	xxv. 3
,,	,,	xxix. 7	,,	,,	xxv. 10
,,	,,	xxix. 12	338	,,	xxv. 42
299	Deut.	xvi. 1	,,	Dan.	i. 1, 3, 4, 6, 7, 17–20
,,	I. Kings	vi. 1, 37	340	,,	viii. 13, 14
,,	,,	vi. 38	,,	,,	xii. 7
,,	,,	viii. 2	,,	,,	vii. 25
300	Esther	ii. 16	,,	Rev.	xii. 14
,,	,,	iii 7, 13	341	,,	xiii. 5
,,	,,	viii. 9, 12	,,	,,	xi. 2, 8
,,	,,	ix. 1, 17, 19, 21	,,	,,	xii. 6
,,	Ezra	vi. 15	,,	Dan.	xi. 13 (margin)
,,	Neh.	i. 1	,,	,,	iv. 16
,,	,,	ii. 1	,,	,,	iii. 16–18
,,	Zech.	vii. 1	348	,,	x. 12
,,	Deut.	xxi. 13 (yerach)	353	Josh.	iv. 19
,,	II. Kings	xv. 13	355	,,	v. 10
,,	Gen.	xxix. 14 (chodesh)	,,	,,	vii. 2–5
301	Num.	xi. 18–20, 31 ,,	,,	,,	vii. 1, 21
,,	Psalm	lxxviii. 27	,,	,,	viii.
302	Gen.	vii. 11	356	,,	viii. 30–35
,,	,,	viii. 3, 4	,,	Exod.	xix. 1, 11
304	Ecclus.	xliii. 6, 7	362	Josh.	x. 13
,,	Psalm	civ. 19	360	Luke	ii. 44
308	Exod.	xii. 2	371	Josh.	x. 8
309	I. Chron.	xii. 15	373	,,	x. 10
,,	Jer.	xxxvi. 22, 23	374	,,	x. 12
,,	Ezra	x. 9	375	,,	x. 11
310	Neh.	i. 1, 2	376	,,	x. 27 (R.V.)
,,	,,	ii. 1	378	,,	x. 13
,,	,,	viii. 14	382	,,	x. 13
311	Exod.	xxiii. 16	384	,,	x. 13, 14
,,	,,	xxxiv. 22	385	II. Kings	xx. 5–11
,,	II. Chron.	xxiv. 23	386	Isaiah	xxxviii. 8
312	II. Sam.	xi. 1	387	II. Chron.	xxxii. 31
,,	I. Chron.	xx. 1	389	Isaiah	xxxviii. 8
,,	I. Kings	xx. 26	390	II. Kings	xx. 9 (R.V.)
,,	II. Chron.	xxxvi. 10	,,	I. Kings	x. 5
313	Exod.	xii. 2	,,	I. Chron.	xxvi. 16
,,	,,	xxiii. 16	,,	II. Kings	xvi. 18 (R.V.)
,,	,,	xxxiv. 22	391	Neh.	xii. 37
321	Gen.	i. 5	392	Isaiah	xxxviii. 10, 11
322	,,	vii. 11	,,	,,	xxxviii. 19, 20
,,	,,	viii. 13, 14	393	Matt.	ii. 2, 5–10
325	,,	viii. 22	396	,,	ii. 10
,,	Psalm	lxv. 9–11 (R.V.)	399	Luke	ii. 11
326	Exod.	xxi. 2	400	John	xvii. 3
,,	,,	xxiii. 10, 11			
327	Lev.	xxv. 20–22			

INDEX

	PAGE
ABEN EZRA, Rabbi	260, 278, 305
Abib (month of green ears)	299
Acronical rising	223, 246, 261
Adar, month	85, 300, 304
Aerolites	111, 112, 113
Ahaz, Dial of	385-392
Alexandria, Museum of	5, 6, 138, 139, 290
Algebar, star-name	233, 234
Allen, R. H.	221 222
"Alroy"	221
Aratus	149, 150, 152, 154, 162, 163, 186, 208, 218, 222, 224
Arcturus (see 'Ash)	258-266
Aristotle	76, 105
'Ash	214, 215, 216, 243, 251, 258, 259, 260, 261, 264-266
Asherah ("groves")	67, 88
Ashtoreth	67, 88, 89, 90, 131
Astrology	5, 77, 78, 130-145, 248
Astruc, Jean	171, 172
Atmospheric circulation	41-45
Aurora Borealis	117
'Ayish, see 'Ash	
Baal, or Bel	67, 89, 131, 176, 178, 210, 253
Bear, the (see Arcturus)	152
Benetna'sh	260
Bethlehem, Star of	393-400
Bosanquet, J. W.	388
,, R. H. M.	315
"Boundary-stones"	153, 154, 198, 318, 320
Bow-star, the	240
Bradley, third Astronomer Royal	96
Bul, month	85, 299
Burton, Lady	379
"Canterbury Tales"	277
Cardinal points	50, 51
Carrington, R.	220
Causality, Law of	15, 16, 18, 78
"Chaldean Account of Genesis"	27

INDEX

	PAGE
Cherubim	166, 169, 188, 190
Cheyne, Dr.	238, 240, 254, 255, 256
Chisleu, month	85, 238, 300, 304, 310
Chiun	133, 134, 144
Clouds	42, 43, 44, 46, 54
,, , the balancings of the	46
,, , the spreadings of	49
Colures, the	159
Comets	103–108
,, , Donati's	105, 107
,, , Halley's	103, 104
Conder, Col. C. R.	238
Constellations, list of	151–152
,, , origin of	149–161
Copernicus	76
Cowell, P. H.	303
Creation	12–24
,, , story of, Babylonian	26, 170, 178, 240, 242, 246, 252
,, , ,, ,, , Hebrew	25
,, , ,, ,, , Scandinavian	29
Cycles, Astronomical, of Daniel	337–348
Cylinder seal	71, 217
Damascius	26, 27
Daniel, Cycles of	337–348
Dawson, Dr. W. Bell	343
Day and its divisions	269–282
Days, different kinds of	271, 272
"Dayspring"	71, 281
Decans	142, 244, 245, 248
De Cheseaux	343
Deep (tehōm)	25–34, 53, 201, 210, 211, 234
,, , fountains of	52–54
Delitzsch, Prof. Fr.	31, 157, 170, 171, 285
Deluge	49, 53, 83, 161, 165, 168, 170–185, 254
Denning, W. F.	220
Dial of Ahaz	385–392
Diana of the Ephesians	112
Dieulafoy, Marcel	372
Disraeli	278
Drach	221
Draconic period	122
Dragon's Head and Tail	198, 199
Driver, Dr.	172, 209
Earth (eretz)	39
,, , corners of	51
,, , foundations of	39, 58
,, , pillars of	39, 40
East (kedem, front)	51
,, (mizrach, rising)	51

INDEX

	PAGE
Eclipses	118–129
Edda, prose	29
Ellicott, Andrew	114
Epicureans	71
Epping, Dr.	274
Equuleus	152
Eratosthenes	218
Ethanim, month	85, 299
Eudoxus	5, 6, 37, 152, 345
Euripides	218
Eusebius	88
Evenings, between the two	277–279
"Eyelids of the Morning"	117, 209, 210
"False Dawn"	117
Firmament (*raqia'*)	35–38
,, (*stereoma*)	37
Flamsteed, first Astronomer Royal	96
Flood, *see* Deluge	
Gad	132, 217
Galileo	3, 4, 76
Gamaliel, Rabbon	297
Genesis and the Constellations	162–169
Gesenius	134
Gilgamesh, Epic of	167, 170, 177, 180
Gosse, P. H.	209
Groves, *see* Asherah	
Guinness, Dr. H. Grattan	343
Heaven (*shamayim*)	35, 36, 38
,, , "bisection of"	55, 362
,, , foundations of	39
,, , host of	56, 57, 65
,, , pillars of	39
,, , stories of	40
,, , windows of	49, 50, 53
Heliacal rising	59, 222, 224, 261
Herschel, Sir W.	75, 76
Hershon, P. I.	311
Hesiod	136, 152, 154, 216, 218, 287, 284
Hesperus	137, 232, 258
Hipparchus	5, 96, 250, 345
Höffler, Dr.	266
Hömmel, Dr.	240
Homer	136, 153, 154
Horace	287, 288
Hour (*sha'ah*)	281
,, , double- (*kasbu*)	282, 320, 345, 381
Humboldt	114
Hyades	133, 217

INDEX

Ibrahim ben Ahmed	114
Iliad	80
Istar	90, 131, 253, 323, 324
Jehuda, Rabbi	261
Jensen	240
Josephus	68, 187, 222, 279, 288, 289
Joshua's Long Day	351–384
Jubilee, the	326–336
Jupiter	104, 131, 132, 137, 247, 396
,, (*Nibir*)	243, 247
Juvenal	288
Karaite Jews	278
Kepler	4, 96, 396
Kĕsil	214–216, 231–232, 237–243, 251, 261, 262
Ketu	201
Kimah	214–216, 223, 281–232, 237, 241, 243, 251, 261, 262
King, Dr. L. W.	240, 241, 303
Kouyunjik mound	27, 33
Lance-star	240
Leonid meteors	114, 116
Leviathan	196–212
Longfellow	233, 236
Lucifer	132
Mädler	220
Maestlin	219
Mazzaroth (*or* Mazzaloth)	130, 214, 243–257, 270, 280
Meni	132, 217
Mercury	131, 137
Merodach	28, 29, 33, 131, 167, 178, 210, 234–242, 247, 252
Meteors	111–117
Metonic Cycle	306, 335, 336, 339, 344
Milton	107
Mishna, the	297, 311
Mithraic cult	160
Month	293–304
,, anomalistic	342
Months, Hebrew names for	304
Moon	79–92
,, blindness	92
,, -god (Sin)	87, 253, 323, 824
,, , harvest	81
,, , new	123
,, , phases of	80, 91
Müller, Otfried	262
Newton	4
Nisan, month	300, 304, 310, 311, 315, 820

INDEX 409

	PAGE
Node	121, 122
North (*mezarim*)	262, 263
,, (*tsaphon*)	51
Onias	68
Orion	231–242
Ovid	288
Palestine Exploration Fund, map	360, 362
Panyasis	152
Parallax	73, 98, 265
Persius	288
Peschitta	259, 261
Philo	289
Phosphorus	132, 137
Pinches, T. G.	27, 28, 30, 31, 90, 176, 235
Pleiades	133, 152, 213–230
Precession	158
Pritchard, Prof. C.	397
Proctor, R. A.	107, 108, 135, 141
Procyon	152, 240
Ptolemy, Claudius	5, 76, 96, 149–154
Ptolemy Philometer	68
Pythagoras	137, 345
Rahab (the proud one)	204–206, 211
Rahu	201
Rain	42–45, 49
"Records of the Past"	26, 28
Remphan	133, 134
Ring with wings	88, 126, 129
Ruskin	46
Sabbath	22–24, 283–292
Sabbatic Year and the Jubilee	326–336
Samaritans	278
Samas (sun-god), *see* Sun	
Sanchoniathon	88
Sanhedrim	296
Saros, the	122, 123, 346
Saturn and Astrology	130–145
Sayce, A. H.	33, 315
Schiaparelli, G. V.	7, 41, 43, 139, 145, 198, 253, 254, 261–263, 269, 279, 285, 286, 290
Septuagint Version	37, 133, 134, 161, 215, 231, 241, 258, 259
Sin (moon-god), *see* Moon	
Sirius	98, 240
Sivan, month	303, 320
Smith, George	27, 30
South (*darom*, bright)	51
,, (*negeb*, desert)	51

INDEX

	PAGE
Star of Bethlehem	393–400
Stars	75, 95–100
,, , morning	59–61
,, , royal	160
,, , shooting	113
,, , Triad of	253, 320
Statius	222
Stern, Prof.	261, 262
Strassmaier	274, 285
Sun	55, 63–78
,, -god (Samas)	67, 131, 174, 253, 323, 324
,, -stroke	72
Talmud	222, 279, 297, 311
Tammuz	66
Targum, the Jerusalem	190
Tavthé, *see* Tiamat	
Tehōm, *see* Deep	
Tennyson	36, 79, 80
Thales	345
Thiele, Prof.	15, 16
Tiamat, or Tiamtu	27–29, 32, 34, 201, 210, 234–235, 240–242
Tibullus	288
Tides	41, 53, 92
Tribes of Israel and the Zodiac	186–195
Tycho Brahé	96
Venus	90, 131, 132, 136, 137
Virgil	160
Vulgate	258, 259
Week and the Sabbath	283–292
West (*mebō hasshemesh*, going down of the sun)	51
,, (*yam*, the sea)	51
Winckler, Prof. H.	235
Winds	50, 51
Wormwood, the star	116
Xenophanes	71
Year	305–325
,, (*shanah*)	305
Yehoshua, Rabbi	297
Zeuchros	142, 249
Zif, month	85, 299
Zodiac, constellations of	141, 151, 152
,, , sections of (*mizrata*)	243, 251
,, , signs of	141, 245, 249
Zodiacal Light	117

www.ingramcontent.com/pod-product-compliance
Lightning Source LLC
Chambersburg PA
CBHW030214170426
43201CB00006B/78